SKETCHING FOR ENGINEERS AND ARCHITECTS

Using real working drawings from a fifty-year career, Ron Slade shows how drawing remains at the heart of the design process in the everyday working life of engineers and architects. The book explains simple techniques that can be learnt and used to enhance any professional's natural ability. Using over 180 categorised examples, it demonstrates that drawing remains the fastest, clearest and most effective means of design communication. Unlike many other books on drawing in the construction industry, this book is 'engineer led' and science oriented, but effectively shows that there is a close affinity between the working methods of architects and engineers.

Ron Slade is Structural Director at WSP | PB Group. Ron received his B.Sc. First Class Honours in Civil Engineering at City University, London and became a chartered member of the Institution of Structural Engineers in 1971 when he was awarded the Institution's A. E. Wynn prize. He was first appointed as a director in 1982.

'Good engineers think, design and communicate through their sketches. A thoughtfully hand-drawn sketch offers a wonderfully efficient and immediately satisfying way for expressing the core concepts of a design. In fact, many problems and solutions do not reveal themselves until drawings are made from different viewpoints. The very act of drawing can help clarify the fabrication sequence and constructability of a complex design. Ron's wonderful sketches are a delight to the eye and the mind. I can think of no better recommendation to my undergraduate and postgraduate structural engineers than that they obtain a copy and cherish this delightful book.'

—**Roger Crouch, Professor and Dean, School of Mathematics, Computer Science and Engineering, City University, London, UK**

'Busy with 21st century technology, we run the risk of losing our mother-tongue: Sketching. Ron Slade's book *Sketching for Engineers and Architects* is a must-have for all aspiring design and construction leaders in the building industry. This book is a treasure chest overflowing with creative engineering sketches and easy-to-understand drawing concepts. We are inspired and patiently guided to set aside our computers more often and pick up our pencils to organize, explore and communicate our ideas.'

—**Gregory Brooks, Senior Lecturer, Architectural Engineering, The University of Texas at Austin, USA**

'In a world of 3D modelling, the skill of interpretation through drawing is being lost. CAD has given us the ability to model buildings virtually, but can never replace the skill of engineers like Ron in being able to truly understand the challenges through free-hand construction sketches showing the process from fabrication through to construction and in doing so, developing innovative solutions. His sketches remind us of the importance of embracing technology whilst recognizing the role traditional methods can play in successful engineering.'

—**Peter Miller, Sales Associate Director, Severfield, UK**

'In this age of digital imaging, 3D modelling and all manner of computer-aided drafting I believe that the art or skill of sketching is as valuable and effective a means of communicating an idea as any modern communication media. Sketching is a crucial tool in the kit of anyone who is engaged in the design or engineering process and should rank alongside IT in the education and development of young aspirant designers.'

—**Peter Emerson, Laing O'Rourke, UK**

SKETCHING FOR ENGINEERS AND ARCHITECTS

Ron Slade

Routledge
Taylor & Francis Group

LONDON AND NEW YORK

First published 2016
by Routledge

2 Park Square, Milton Park, Abingdon, Oxfordshire OX14 4RN
52 Vanderbilt Avenue, New York, NY 10017

Routledge is an imprint of the Taylor & Francis Group, an informa business

First issued in hardback 2019

British Library Cataloguing-in-Publication Data
A catalogue record for this book is available from the British Library

Library of Congress Cataloging-in-Publication Data
Names: Slade, Ron, author.
Title: Sketching for engineers and architects / Ron Slade.
Description: New York : Routledge, 2016. | Includes bibliographical
references and index.
Identifiers: LCCN 2015044453| ISBN 9781138925403 (hardback : alk. paper) |
ISBN 9781315683775 (ebook)
Subjects: LCSH: Architectural drawing—Technique. | Freehand technical
sketching—Technique.
Classification: LCC NA2708 .S58 2016 | DDC 720.28/4—dc23
LC record available at http://lccn.loc.gov/2015044453

ISBN: 978-1-138-92540-3 (hbk)

Typeset in Arial
by Florence Production Ltd, Stoodleigh, Devon, UK

CONTENTS

FOREWORD

I was delighted to be asked to add a foreword to Ron Slade's book as I have long been an admirer of his engineering prowess, particularly as expressed in his exemplary drawings. Like me, he is a person who feels most comfortable expressing himself visually, but the great strength of his drawings and his approach to visual language is that it is rarely the drawing in itself that is the driver. It is the drawing as an act of communication, of thought processes and solutions to problems that is the primary value. So it is drawing as a language tool, as part of the process of 'problem-solving' that answers the need to explore and describe answers to the very practical and pragmatic demands that architects and engineers address on all construction projects. I particularly enjoy Ron's work when it embraces time-related sequences, as it's often not the final answer that is paramount, but how you get there, how to lead up to and assemble; the process of constructing the actual solution can often be every bit as important as the very solution itself.

This book, in an age of increasing reliance on computer based drawing, is a very timely reminder of the crucial ongoing importance of the immediacy and spontaneous communication of hand drawing. But what comes over again and again is Ron's inherent modesty and his desire to pass on his skills generously to young architects and engineers. He comes from several generations of passing on skills, as grandfather and great grandfather were stonemasons – a trade whose skill sets often led to the very first architects and engineers; and his father, a builder who no doubt imbued the young Ron with yet further breadth, encompassing a wider view of construction, and the processes of assembly and actual on- and off-site organisation.

A further trait of Ron's is his consistency of approach that I and so many architects such as Michael Hopkins, Renzo Piano, SOM, Foster + Partners, KPF, my former partner Nick Grimshaw and others have continuously relied on over many decades and on many notable projects. And this consistency, integrity and reliability is borne out right from the start of his career by never changing jobs – he has forever been with WSP | PB (and its merged partners Kenchington Little) throughout a long career that has seen such vast, sweeping changes in our industry – not least being growth and size. Kenchington Little when he joined was around 100 employees and WSP | PB is now 31,000 – one of the largest organisations of its kind in the world.

What makes this book important today is that it maintains and promotes faith in the basics that have and will stay consistent in spite of all around changing in our ever busier and more complex world. Hand drawing, the ability to communicate concepts and complex processes visually, is an art, a skill that can be developed, but some, like Ron, are touched with genius in their attainment of a higher order of achievement. He, like his best students, is at core an enthusiast for whom his profession is more than a job. For young people embarking on a wide range of careers in the construction industry this is a welcome primer that advocates and explains the ever continuing value of hand drawing in designing and putting up buildings.

Sir Terry Farrell
Autumn 2014

PART 1: INTRODUCTION

Looking back over fifty years spent working in a consulting engineer's office, the changes that have affected day-to-day practice are amazing. Our ability to understand materials and the behaviour and performance of complex structures has increased. Engineers no longer rely on slide rules and log tables but on sophisticated analysis software. The basis of design has moved from a simple allowable stress approach to limit state design, an approach where criteria are set such that within acceptable probability, a structure will not reach a limit state in which it fails or is in some way unserviceable. This was a profound and far reaching change in design procedures. Organisations also changed and became more overtly business led and not surprisingly, given the advances in IT, the methods of conveying design intent changed radically. The last drawing boards disappeared from the office 15 or 20 years ago. Now we use CAD and we exchange digital information, we produce 3D models and we can print 3D models.

One thing hasn't changed. Many engineering designers are employed by organisations that still call themselves consulting engineers. The use of the term 'consulting engineer' is appropriate as these firms continue to act in an advisory capacity on professional matters. They employ specialists who give expert advice and information but essential to their *modus operandi* is the ability to communicate. In engineering and related industries, this entails providing or receiving information, exchanging views and ideas with a wide range of people including clients, designers in other disciplines, contractors and skilled building workers.

However, despite the advances in computer technology and the exchange of electronic data, simple hand sketching and drawing still has its role to play in fast and effective communication – to get a thought on paper is often a first step in a design process. Pencils have not disappeared from the office. Consulting engineering or engineering in general can never become a totally electronic profession populated by analysts and modellers.

Good design and successful buildings owe much to the effective communication in particular between engineers and architects. Both professions need people who are good communicators. Although this book is written from an engineer's perspective and does not address the artistic quality of drawing and sketching in any depth, the basic approach, ideas and techniques are relevant to both engineers and architects. Hand sketching is a practice that has always had the potential to bring the two cultures of engineering and architecture together.

PART 2: DRAWING AND SKETCHING

2.1   Drawing and its historical context

The Cave of El Castillo paintings in northern Spain are about 40,000 years old and are probably the oldest in Europe. In Australia and Indonesia, cave paintings of a similar age have been found depicting long extinct fauna. The most common images are of large animals (bison, horses, aurochs and deer), tracings of human hands and abstract patterns and symbols. Art experts believe that the paintings and drawings are not just images, they are not simple pictorial representations of an animal or a human being or a shape – they are embellished and enhanced to convey a more complex or deeper message or to satisfy some aesthetic imperative.

Art historians speculate on the reasons primitive people made drawings but most were almost certainly made for magical, superstitious or religious reasons. Whatever the reason, it is clear that there has always been a belief in the power of picture making, the power of images. With regard to the development of drawing techniques, most experts agree that its history, at least in some parts of the world, can be more easily traced back to the ancient Egyptians rather than the Primitives. The Egyptians developed a long lasting style that was not art for art's sake, not beautification but image making for functional reasons designed to preserve likenesses, life styles, status and hierarchies and to give guidance in the afterlife. Perfect clarity was the objective. The ideas of foreshortening and more lifelike representations were left to later ages.

Egyptian art must have been handed down from master to pupil and hardly changed for 3,000 years until the Greeks came into closer contact with the Egyptians. The Greeks moved things on, tackled foreshortening for instance, and all later civilisations learnt from the Greeks, especially after the 'Great Awakening' when science, art and literature made great strides forward.

The aesthetic quality of drawings owes much to the designer's appreciation of proportion and geometry and does not just happen by accident. Rules for building in particular, came from the Ancients. The English *The Constitutions of Masonry*, written in the fourteenth century, opens with the words: 'Here begin the Constitutions of the art of Geometry according to Euclid.' Later we are told that Euclid taught the art of masonry through geometry to the Egyptians. These skills were apparently passed to the captive Israelites and eventually through David and Solomon to King Charles the Second of France. From there, according to the *Constitutions*, they were taken through St Alban to King Athelstan of England, whose son became a master of masonry, and founded the professional organisation of masonry builders and architects. An earlier document entitled *The Old Book of Charges* gives a similar account of how the geometrical art of Euclid was passed from Egypt to England. Medieval masons clearly understood the science of geometry used in architectural design and construction which came from its inventor, Euclid. Although the geometry used by medieval architects was traditionally ascribed to Euclid and his postulates,

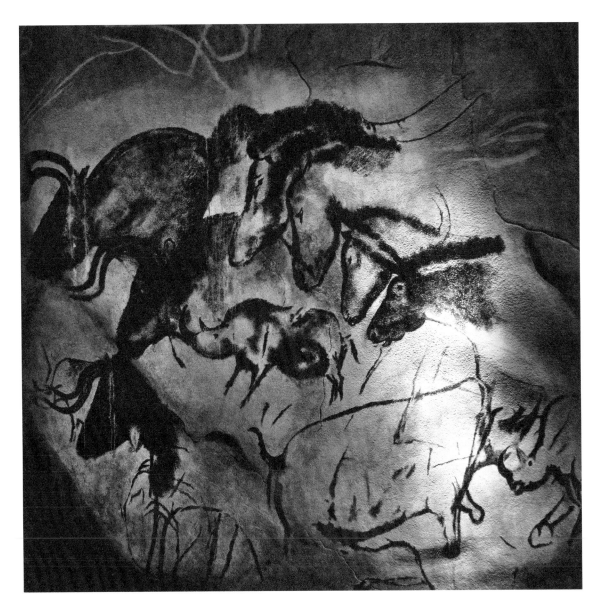

FIGURE 2.1
El Castillo cave paintings in northern Spain
Source: 'Art and Design, the Guardian'.

the architects and masons would have had no direct knowledge of Euclid's works apart from example and craft skills handed down during their apprenticeships.

The fifteenth century saw the beginnings of the Renaissance and the rediscovery of classical art and the new role of the artist as a professional. This led to an expansion in the concept of drawing, and instead of being just a craft skill, drawing became a tool for investigating the natural world, and gave artists greater ability to express their own views of the world around them. Drawing became a tool for design and experiment, and with the dawn of a system to describe the three-dimensional world – linear perspective – the boundaries of drawing expanded enormously.

It is impossible to ignore Leonardo da Vinci (1452–1519) – painter, sculptor, musician, mathematician, inventor, anatomist, geologist, cartographer, botanist, writer and of course architect and engineer. His journals contain over 13,000 pages of notes and drawings and the sheer quality of his sketches is beyond question. Two thoughts for us to remember: first, he collaborated with others, physicians for example with regard to his anatomical drawings, very similar to the collaboration between engineers, architects and other professions today; and second, his use of annotation, which of course separates his scientific drawings from his magnificent artistic works. Leaving room for annotation is important.

In his *A Short Book About Drawing*, Andrew Marr says: 'Drawing had always been important for architects but as science began to advance, it became an important skill for mathematicians, anatomists, collectors of botanical rarities, designers of military fortresses, astronomers.' He goes on to say that suddenly more and more people were sketching 'as a kind of quick verbal information' and that in the 1700s, 'the real drawing craze spreads from small numbers of enthusiasts to the new middle classes' and that some of this continues 'for basic practical reasons'.

Throughout the nineteenth and twentieth centuries a great deal of effort was devoted to attempts to discover the true nature of the geometry used by medieval architects. However, little true geometry was discovered. In John Harvey's words:

> Discussion has been obscured rather than clarified by a vast literature concerned with a possible symbolism, numerology (a little of which was probably intentional), and arithmetical and algebraic analysis. Almost the whole of this literature must be disregarded in seeking for the empirical means by which the architects reached their remarkable results.

In general terms, when produced with care and thoughtfulness, engineering or architectural drawings are so often the best way to communicate and develop ideas, thoughts, intentions, visions, ambitions, commitment and even passion and belief. They become much more than just lines on paper and convey a message more easily than words alone. It is also evident that well thought out drawings develop an intrinsic aesthetic value of their own which we may recognise subliminally but may not appreciate consciously unless we stop and look for it. They just have to feel 'right'. Ideograms, symbols or signs used in writing systems in China for instance, directly represented a concept or thing, rather than words. Although we may not understand the meaning of some of these drawings and symbols, their aesthetic quality has

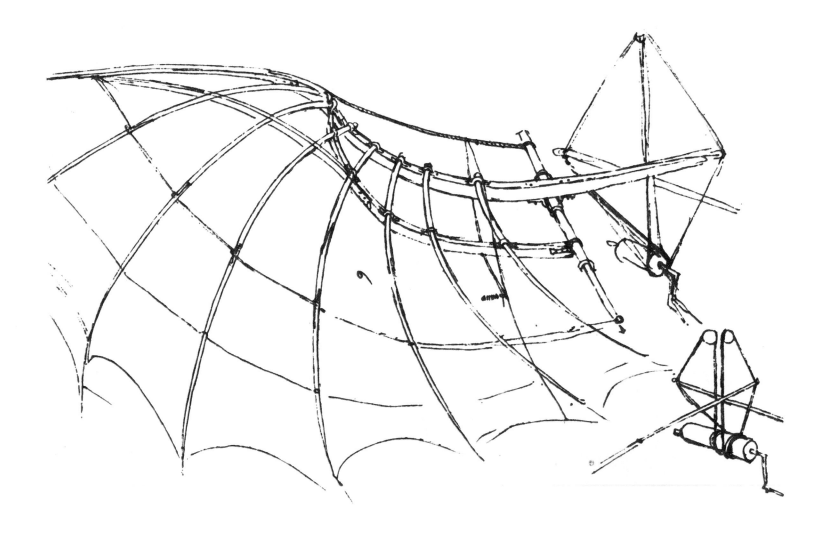

FIGURE 2.2
Flying apparatus, Leonardo da Vinci

ensured their survival from prehistoric times and through the ages to the present day. Can anyone doubt the simplicity and beauty of stylistic ancient Egyptian images, that they were full of meaning? They must have given great satisfaction to the artisans who produced them, and to countless people who have set eyes on them since. The hand drawings of many of our leading engineers and architects are just as thoughtful and appealing. In broad terms, whether in art, architecture or construction, and despite the constraints of rule makers, drawing has always been used with a great deal of freedom for a whole range of purposes, from how to build a building to the expression of personal feelings.

Above all, drawing is about communication and is universal. Throughout history drawing has remained a constant reason for picking up a mark-making implement. It has survived the introduction of the camera and will survive the computer screen. The benefits of drawing and hand sketching should not be discounted or under-valued.

*

2.2 Why sketch in an age of computer generated images?

Figures 2.3 and 2.4 show trees around a pond, and convey the imagined or extant scene in a perfectly clear way but without the benefit of isometric or perspective drawing.

These drawings have been made with great care and each has been made for a purpose. They are beautiful and meaningful and demonstrate the skill involved in image making.

But why use hand sketching to record new ideas in this day and age? The invention and popular dissemination of photography has had a fundamental effect on artists' drawing. The need to copy 'reality' has reduced; the camera can be used instead. Yet drawing still has its vital role to play as a way of thinking, developing ideas quickly, explaining and communicating. The drawing will always out-perform the photograph.

Take for example the work of field archaeologists. Egyptologist Mary Hartley of Macquarie University, Sydney says

> Archaeologists who publish their reports have many of the artefacts drawn by hand. We have an established way of doing things, because we find it gives the most informative results. The drawings are first completed on site in pencil and later inked for publication. Photographs are very important, but the detail seen with the naked eye, reproduced in the drawing, can often add important information to the final illustration. The tiny details are very important, and the sharp black

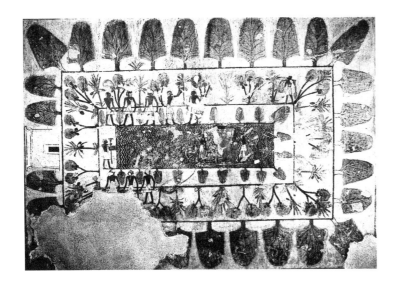

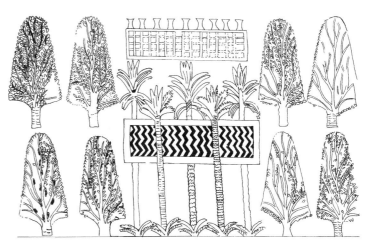

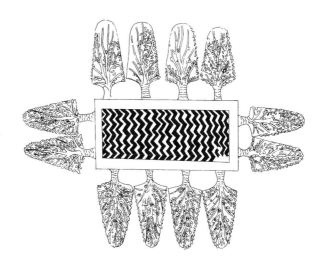

FIGURE 2.3
Pond in Rekhmire
Source: By kind permission of Leonie Donovan.

FIGURE 2 4
Rehkmire: two aspective ponds
Source: Drawing by Mary Hartley after N. de Garis Davies (1953), *The Tomb of Rekh-mi-Re at Thebes*.

and white of the final inked image is clear and easy to understand. All the details are reproduced using a 'dotting' technique, which produces a three dimensional effect. So by including the drawings, the inkings and the photographs, the visual records are complete. [See Figure 2.5.] Hand pencil drawings are particularly effective in illustrating pottery. We have a set way of drawing a pot, dividing it in half – one half indicates the detail found on the outside, and the other half indicates the profile – something a camera cannot do. [See Figures 2.6 and 2.7.]

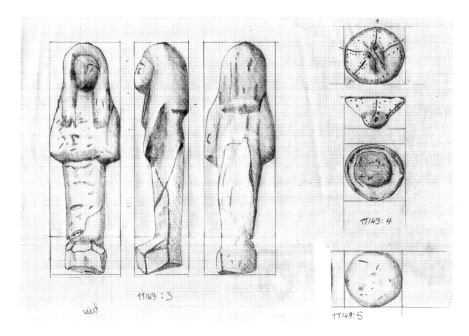

FIGURE 2.5
A page of shabtis according to their typology
Source: Drawing by Mary Hartley from Macquarie University's Theban Tombs Project, TT149, Dra Abu el Naga, Luxor.

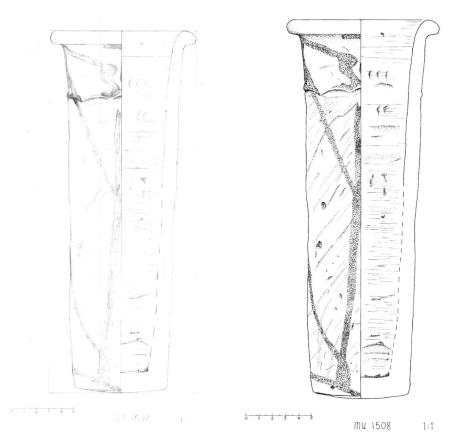

FIGURES 2.6 AND 2.7
Drawings by Mary Hartley of pre-dynastic decorated vessel belonging to the Museum of Ancient Cultures, Macquarie University, Sydney. 2.6: field drawing; 2.7: final inking for publication.
Source: Drawing by Mary Hartley after N. de Garis Davies (1903), *The Rock Tombs of El Amarna* (London), p. 40.

Mary makes another interesting point. She says ancient Egyptian aspective or 'flat art' drawings found on tomb walls illustrate what your eye actually cannot see! Sections and frontal views are often combined and are used in conjunction with the principle of transparency. For example Figure 2.8 shows at the side what is stored in the rooms. Figure 2.8 also shows how a modern architect would draw the building from the aspective plan above. The layout is given in plan; however, the gateways, doors and pillared porticos are added in frontal, even if this meant turning them at right angles to each other.

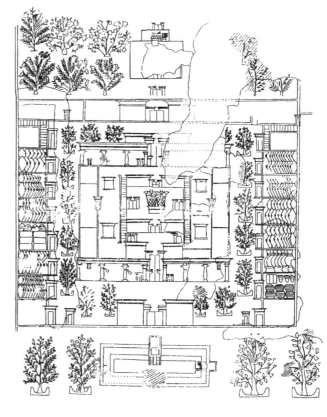

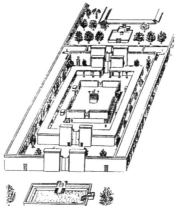

FIGURE 2.8

Amarna 1: aspective plan of palace

Source: Drawing by Mary Hartley after N. de Garis Davies (1903), *The Rock Tombs of El Amarna* (London), Pl. XXXII.

Botanical illustration is another interesting example. All good botanical field guides and reference books are based on detailed and accurate drawings and paintings and not on photographs. The books are clearer, more helpful and informative and are better for recognition and identification. What isn't shown is also important; extraneous material that might exist in a photograph is eliminated – drawing noise is omitted. The illustrations are valued works of art in their own right. Take for example Julie Small's illustration of *Streptocarpus primulifolius* (Figure 2.9). Although hers are far beyond the skills of most of us, to understand her reasons for choosing to invest in hand drawing is revealing.

In Julie's words:

> I feel that drawing in pencil allows the artist to depict a variety of textures, tonal depth and small detail that it is not possible to obtain with a photograph. In order to complete a botanical study I use a number of grades of pencil to create an image, and it is very satisfying to see the subject 'coming to life'. Above all, the longer you look at a subject, the more you begin to see detail that you did not initially observe and ultimately this then presents the challenge of how to transfer that to paper using just a pencil.

FIGURE 2.9
Streptocarpus primulifolius
Source: Julie Small, SBA.

Professor Gregory Brooks of the University of Texas gives another explanation of the value of sketching. He says sketching is a language which has positive effects on memory so becomes a valuable teaching tool:

> The first thing that I do with my new architectural engineering students at the university is teach them to sketch. I explain to them that design studio is in fact a foreign language course. Sketching will be their language, which can then progress into more advanced methods of visual communication, but sketching is the fundamental and universally understood language of design. Without the ability to sketch we are powerless to solve and explain ideas – as designers, we are mute unless we can sketch.

> And so the first week of design studio is spent sketching. We begin by sketching from slides, learning to see. An amazing thing happens almost immediately with the students: rather than merely looking at an object/image, they learn to truly see it. In order to sketch, they must study the object, for they are re-constructing it. Even a simple object must be examined for its shape, proportion. We begin by sketching very lightly, barely touching the paper, and examining the outline for correct proportion. Once the proportion is correct, we can commit with more visible line-weight. Correct proportion is the key, because of course, if the initial proportion of the object is incorrect, any further time spent on that sketch is a lost cause.

> Once the students can recognize and translate correct proportion and form in their sketch, they see the world differently, and anything sketched is ingrained in their memory. One of my favorite exercises is to sketch an object from a slide, critique it together to make adjustments, sketch it again, but much faster, and finally – a day later when the students have long 'forgotten' that sketch – I ask them to sketch the object from memory. They quickly sketch the object without seeing it. I then turn on the projector and show them the slide – they have all invariably drawn it correctly – the object has become part of their memory. This always has a great impression on them and on me.

Professor Roger Crouch, Dean, School of Mathematics, Computer Science and Engineering, City University, London says:

> Good engineers think, design and communicate through their sketches. A thoughtfully hand-drawn sketch offers a wonderfully efficient and immediately satisfying way for expressing the core concepts of a design. In fact, many problems and solutions do not reveal themselves until drawings are made from different viewpoints. Therefore three-dimensional spatial literacy and an ability to produce suitably scaled, proportional sketches is a much valued skill. The very act of drawing can help clarify the fabrication sequence and constructability of a complex design.

These are just three areas where hand drawing has retained its importance: archaeology, botanical illustration and teaching. There are many others of course, ranging from illustration, cartoons and animation, to all areas of design whether fashion or aerospace.

It is clear that sketching and drawing has its place in the world; and it is not surprising that it is an essential tool for both the engineer and the architect. Anything to support the creativity of construction is important – and to construct is to be human – construction is about building things, it is what we are.

Engineering and architecture are 'designing arts' and cannot really be separated – architecture and (civil) engineering were essentially the same discipline characterised by their use of drawings, models and mathematical calculation as a means for constructive forethought. With the proliferation of knowledge, separate discipline streams have had to evolve, but engineering and architecture remain collaborative pursuits and remain inextricably linked to other more specialised professions. These days collaboration must include many other stakeholders, disciplines and specialists and therefore the ability to communicate is more challenging. Sketching and drawing is the oldest, quickest and simplest way of sharing and developing ideas.

So why should engineers and architects sketch, why draw? There are so many reasons:

- It is a way of describing ideas – ideas have to be 'materialised'.
- It can 'physicalise' a concept, it can allow building on paper.
- It facilitates 'thinking' by design development on paper.
- It is a means of exploring ideas.
- It is a way of sharing and developing ideas.
- It promotes collaborative 'face to face' working – sometimes called 'conversational drawing'.
- It can produce unparalleled clarity.
- It can be definite and confident or conceptual and ethereal.
- It provides a human/personal touch.
- It can sometimes circumvent the clinical nature of computer generated images that may ultimately be misleading.

In addition, as a kind of bonus, the process of drawing and sketching can be deeply satisfying and rewarding in itself.

Finally, Gordon Deuce, chief engineer at Mace, has an interesting anecdote. He says early in his career he was advised by his lead engineer: 'Stop calculating and draw it; if you can't draw it, you can't build it.' Many years later he says, it still rings in his ears.

*

2.3   Easy and effective techniques

True freehand

Freehand sketching is so often the method that engineers and architects use to capture emerging ideas. It is the quick and informal way of conveying mental pictures to paper. It often works best 'face to face', engineer and architect, and becomes conversational drawing. It is a form of language, it is often iterative, it promotes lateral thinking, progression and evolution. The scribbled ideas are later developed by a process of refinement and further iteration until a point is reached when it is worth investing more time in sketches that are more rigorous and more carefully thought out.

The earlier in the process you are, the more you need to focus on broad principles and the thicker and softer your pens and pencils need to be. Sketch quickly using the right tracing paper, do not be afraid to make mistakes and go over and over the design ideas until you find a clear route forward. Produce one

sketch after another, keep them as a record if you want or if you are using pencil, keep photocopies at various stages and then use an eraser to redefine your intensions and outlines.

Young engineers in particular sometimes find it difficult to release themselves from the constraints of technical drawing, from the tyranny of the straight line, but it is well worth developing true freehand skills, there is a place for fast freehand. And practice is the key to learning – teachers of life drawing tell

FIGURE 2.10
Rachel

FIGURE 2.11
Rachel

you to use messy charcoal and force you to draw a pose every two minutes. You soon become more proficient.

To my mind it doesn't matter how neat and tidy your early sketches are; after a design session you are more likely to save just a few and throw the others away. Some people can set down their thoughts and make legible sketches in their notebooks during design meetings – see for instance an example of John Parker's work in Figure 2.12 – but I prefer to summarise ideas immediately after the meeting (my notebook sketches are messy) in what we at WSP | PB have come to know as 'design development notes' (Figure 2.13).

There are plenty of textbooks to show you what sketching tools to use, what line thickness, how to draw straight lines, vertical lines, horizontal lines, oblique, parallel and curved lines, but let's move onto the more structured sketching that will in most cases have a longer shelf life.

Guided freehand

Innate skill is not essential in the production of successful sketches or drawings. Getting reasonable results is not difficult, you just have to persevere and have the confidence that it will come good in the end. Keep things simple – choose axonometric projection rather than perspective, perspectives take longer (but if you do use perspective, use Anderson Inge's 'cube method' – see page 18). Isometric and axonometric projections are simple ways to approach 3D drawings. They can be used for sketches or to draw to a predetermined scale, and if a scale is used, every part of the drawing can be measured with

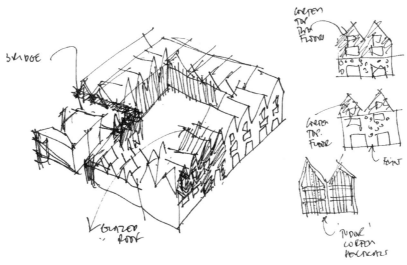

FIGURE 2.12
A page from Alex Lifschutz's note book. (Alex is a director at Lifschutz Davidson Sandilands.)

15

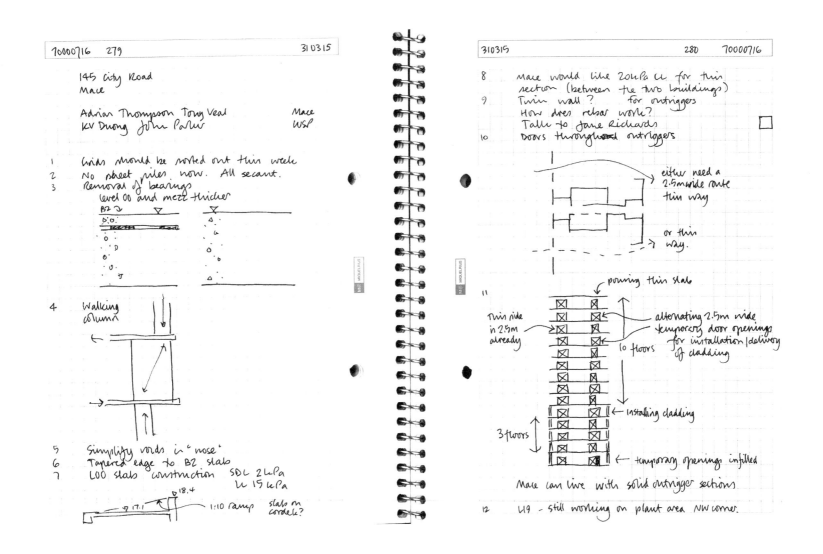

FIGURE 2.13
A page from John Parker's note book

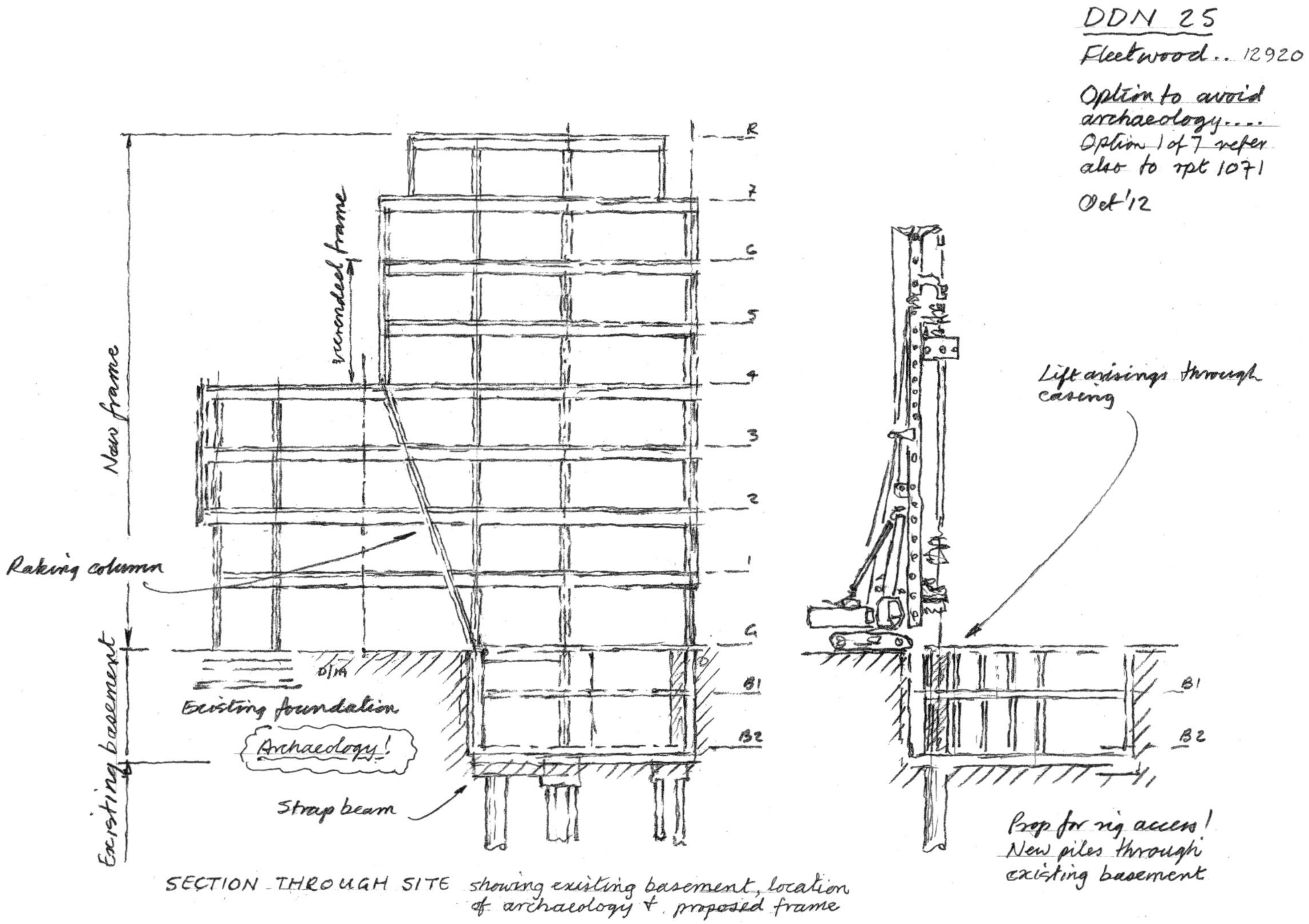
DDN 25
Fleetwood.. 12920
Option to avoid archaeology....
Option 1 of 7 refer also to rpt 1071
Oct '12
New frame
Existing basement
Raking column
Existing foundation
Archaeology!
Strap beam
Lift arisings through casing
Prop for rig access!
New piles through existing basement
SECTION THROUGH SITE showing existing basement, location of archaeology + proposed frame

FIGURE 2.14

Quick sketches and minimal text summarising an option for further development (a design development note)

accuracy. Unlike perspective drawing, lines in isometrics or axonometrics do not converge. In fact they only go in three different directions. Isometric projections use vertical lines and lines 30 degrees to the left and right. In axonometric projections, the lines are vertical and lines on the plan are at 90 degrees to each other. Axonometric projections may look distorted but can appear much more realistic if the vertical lines are drawn to 70 per cent of true scale.

Figure 2.17 shows a page from Anderson Inge's sketchbook demonstrating the cube method he teaches for drawing quick perspectives. The cube method avoids constructing from plan drawings, and it begins with establishing a single correct cube as a base for the building or object that is planned for the page. Adjust the base cube, along with its vanishing points and horizon line, until they look right to the eye. In this example, the Arc de Triomphe has been built-up by extending the base cube into a group of 3×3×1 cubes for the whole monument. The front 'measuring line' can be relied on to maintain accurate scale for the perspective, and diagonal lines are used to extend this scale backwards (or forwards) to give other cubes correct perspective appearance. Once the cubes have been laid out, it is straightforward to introduce edge profiles and surface features. Pay special attention to the diagonal line, it is key to drawing the similar volumes next to each other.

Both types of drawing can be produced quickly by placing gridded paper below tracing paper. This approach ties in with the notion that engineers and architects are practical people and do not necessarily rely on artistic flair. Worthwhile results can be produced by 'constructing' drawings and sketches.

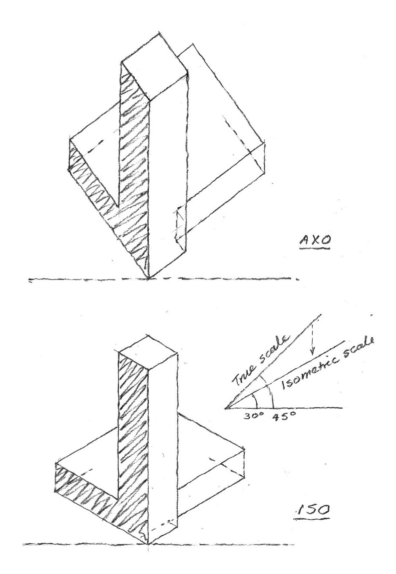

FIGURE 2.15
Axonometric and isometric

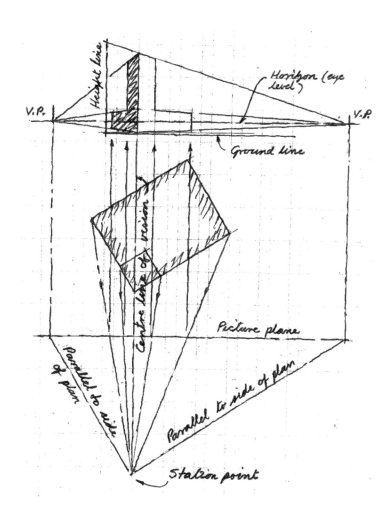

Image labels (as shown in figure):
- Height line
- Horizon (eye level)
- V.P.
- V.P.
- Ground line
- Centre line of vision
- Picture plane
- Parallel to side of plan
- Parallel to side of plan
- Station point

FIGURE 2.16
Two point perspective

Some practical tips:

- Line quality, shading, hatching and composition are hugely important. Just go for consistent line weight and parallel line hatching. Getting composition right is about setting things out on the paper and not squeezing something up into a corner.
- Learn to draw straight lines and parallel lines.
- Draw lines faintly and then 'line in' after checking straightness or curvature.
- 'Feather in' curves – don't expect to produce perfection in one sweep of the pencil.
- Observe and correct mistakes (a pencil in one hand, an eraser in the other).
- Plan your drawing first.
- Decide what to show and what to hide.
- Cut sections – sections make you think.
- Use any tricks you want – trace photos, use a light box, get basics drawn using CAD, use the photocopier and Tippex to develop sequence sketches ... who cares so long as the result is good?
- Develop your own library of common objects that you can re-scale and trace (a tower crane or a piling rig).
- Try 'reversing out' – even a simple sketch looks better that way – see Figure 2.18, a rather architecturally uninspiring design for a high bay warehouse, where the sketch shows construction on two fronts away from laterally stabilised central bays.

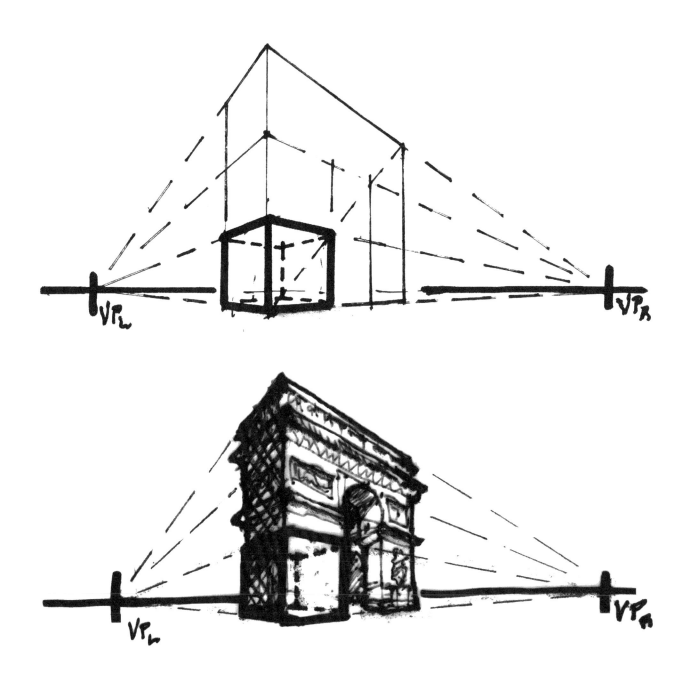

FIGURE 2.17
Anderson Inge's *Cube Method*

FIGURE 2.18
High bay warehouse

I usually draw on tracing paper, preferably 25gsm lightweight sketching paper with a regular 0.7mm automatic pencil using HB leads. Much softer, thicker leads are good for less detailed, broad outline, concept-type work. The paper feels 'right' to me, and yes, it smudges a bit if you are not careful but that doesn't really matter. Buy the paper in a 297mm × 100m roll, much cheaper that way, and get used to drawing and scrapping, and drawing again and throwing away again until you get it right. Then I find photocopying a tracing paper sketch really transforms it, gives you an idea of what needs adjusting to give a satisfying end result.

You will not get things right first time. It is about patience, practice and persistence and it is about putting observations and ideas on paper.

I cannot draw like an artist, I don't have that sort of flair or that kind of imagination. I can, by being doggedly persistent and when time permits, produce reasonable images just by working and reworking until things look right – Figures 2.19 and 2.20.

FIGURE 2.19
Tara

FIGURE 2.20
Corazón Cinco – an illustration for a non-engineering text

Never forget that most people like to draw or sketch, whether doodles or something a little more demanding. One very well-known architect was known to admit that he got into architecture not because he was especially interested in buildings, but simply because he wanted to find a profession where he could draw.

Another anecdote concerns a short office trip to Valencia aimed at discovering the city and studying the works of Santiago Calatrava. After the visit, a group of engineers from WSP | PB's London office put together a book of photographs and sketches with an introductory note from the organiser that really speaks for itself: 'the enjoyment and inspiration that arose from sketching on one's own terms was a welcome surprise to many in the group. People who didn't know they could sketch were sketching away just for fun.'

*

FIGURE 2.21
Sean by Evelina Gadzhova, engineer, WSP | PB Group

ROBERT WIESNER

FIGURE 2.22
Image from the Valencia sketch book

2.4 Axonometric projection

An example

Axonometric projection is the easiest and least complicated form of three-dimensional drawing. Plan and vertical dimensions can be scaled, although if vertical dimensions are reduced by say 70–85 per cent the resulting image looks a little more realistic. Whether vertical dimensions are reduced (foreshortened) or not, the resulting image gives a degree of accuracy missing from pure freehand sketching and avoids all the complex mathematics of true perspective. The example here is intended to show how something can be produced which is recognisable and reasonably accurate. It is produced using aerial and street views available on the internet. The steps are:

Step 1 (Figure 2.23)

- Draw a simple plan based on a street map or an aerial photograph on gridded paper.
- Guess the elevation by looking at proportion (height versus length). Again, draw on gridded paper and with the aid of photographs.

Step 2 (Figure 2.24)

- Trace the plan at a 45-degree angle, or any other angle, it doesn't matter. Choose a convenient mid-height level.
- Position the tracing over gridded paper.
- Measure up and down from your chosen level. Use rectilinear blocks to enclose awkward shapes like curves or domes.

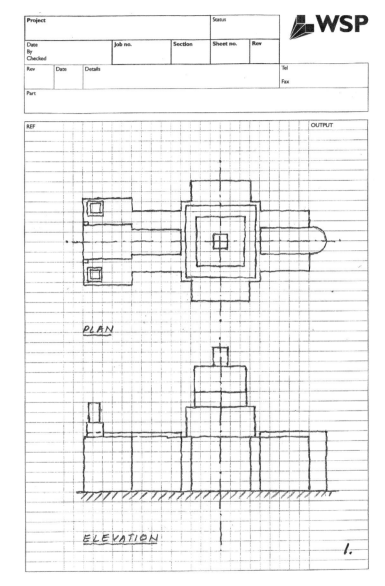

FIGURE 2.23
Axonometric: step 1

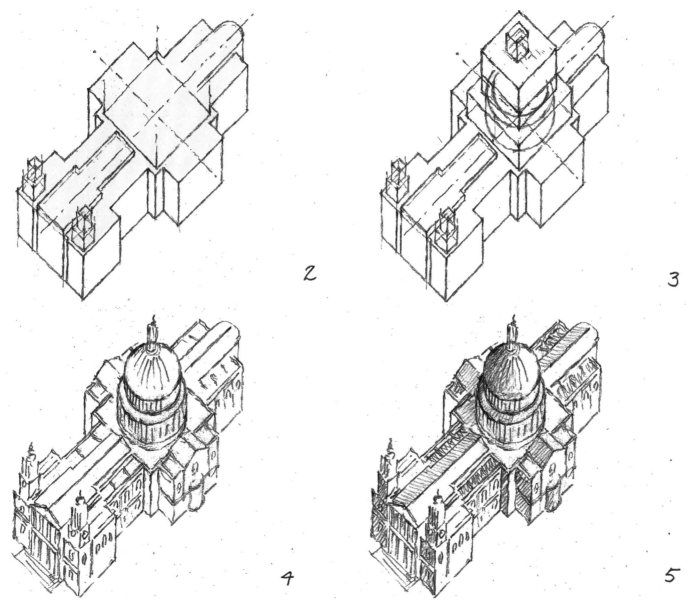

2

3

4

5

FIGURE 2.24
Axonometric: steps 2, 3, 4 and 5

Step 3 (Figure 2.24)

- Complete the 'block' diagram and fit, in this case, the circles into the squares where you will need to draw the dome.

Steps 4 and 5 (Figure 2.24)

- Trace the whole thing again adding shading.

A second example

Axonometric projection again, this time to unravel a mechanical engineer's drawing. First print the drawing off to a scale which will allow your sketch to fit onto, say, A4 sheets of tracing paper.

Step 1 (Figure 2.25)

- Ask questions about the conventional drawing. So often the person drawing the plan and elevation doesn't see the ambiguities – that's a good reason for drawing the 3D sketch in the first place. If there are ambiguities, you won't be able to draw the complete thing, until your questions are answered.

Step 2 (Figure 2.26)

- Assuming you can get the answers to your questions, it's often a good idea to identify separate systems (if they exist) – colour or shading helps, here it's supply and return air from air handling plant. The lower system is not shown coloured but instead is stippled.

Step 3 (Figure 2.27)

- Draw each system separately, using a sheet of gridded paper to draw guide lines to keep lines parallel, whether vertical or the angled orthogonal lines used for the plan view. Use a reference line so that the systems can be related to each other.

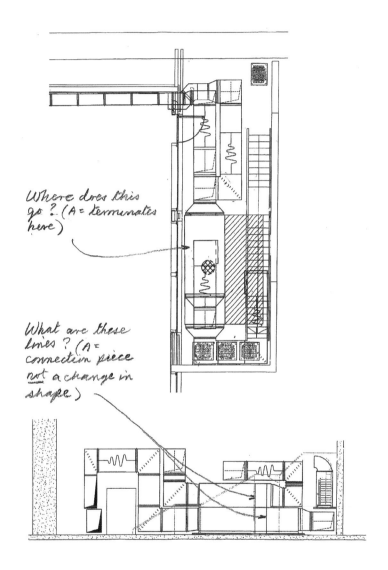

FIGURE 2.25
Axonometric: second example step 1

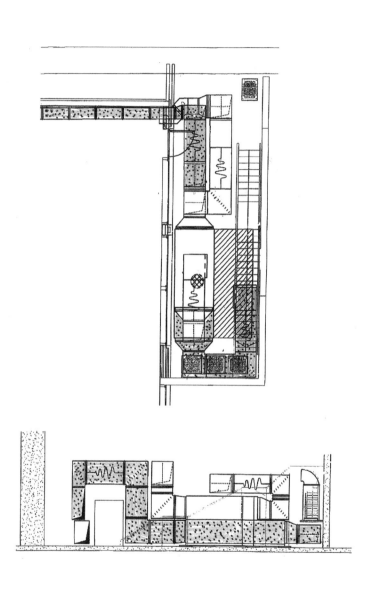

FIGURE 2.26
Axonometric: second example step 2

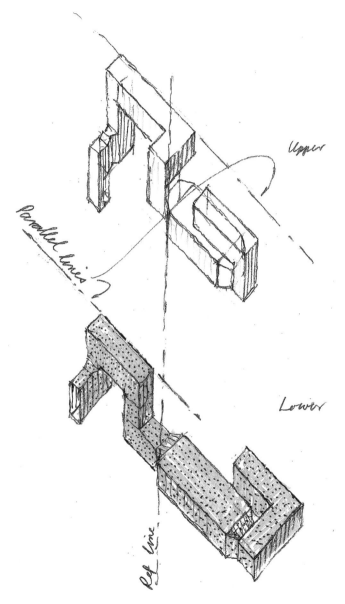

FIGURE 2.27
Axonometric: second example step 3

Step 4 (Figure 2.28)

- Reassemble the two systems, removing hidden lines by re-tracing your original sketches.

Step 5 (Figure 2.29)

- Add supplementary information, the supporting platform, access stair, etc.

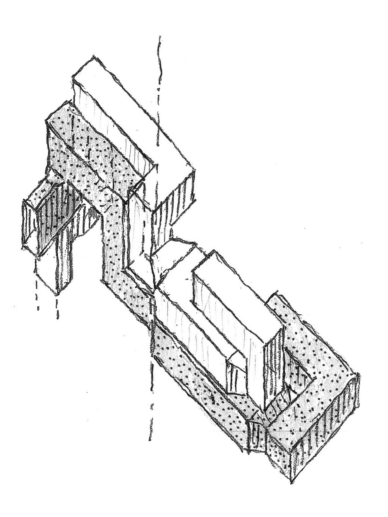

FIGURE 2.28
Axonometric: second example step 4

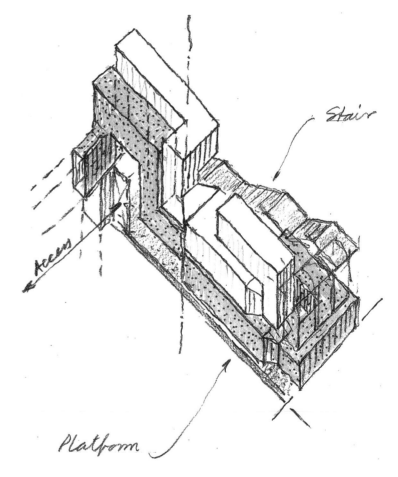

FIGURE 2.29
Axonometric: second example step 5

Draughtsmanship

Some individuals have special talent when it comes to drawing – don't be put off, just appreciate their work and recognise that observation, composition, line quality, patience and perseverance are important to them as they are to successful engineering or architectural sketching and drawing.

Bridging two worlds – pencil sketch and the tablet computer

The first successful tablet computer with a multi-touch screen, the Apple iPad, was released in 2010 and since then, these devices have become ever more sophisticated. Is it possible to transfer sketching skills from paper to electronics? Laurie Chetwood's superb tablet drawings would suggest that it is; he would say it's just another medium and equivalent to moving between charcoal and pencil or between watercolours and oils. Laurie certainly maintains his unique style and his amazing skills are still very evident. It's probably a generation thing, but I have a lingering concern that many less skilful images rely on clever hidden software which produces special effects at the touch of a button – they feel ephemeral to me and almost too perfect. Conversely, pencil hand sketches often provide physical evidence of the thought and effort that goes into their production; you might see guide lines, reworking and even corrected errors and smudges. And I think paper is somehow more personal and tangible – for me it's the difference between a Kindle and a paperback.

FIGURE 2.30
Chinese landscape
Source: Laurie Chetwood (Chetwood Associates).

But let Laurie explain his views in his own words:

Like most things, being successful at something comes from practice and the inclination to practise usually comes from enjoying something in the first place. Being a successful artist is no different. Drawing skills are available to anybody if they feel inclined to practise – it is the desire to pick up a pencil that makes an artist. The ideas which come from these skills are a delineator and once in place – and with them confidence in the outcome – it is possible for a draftsperson to use any of the tools at his or her disposal. The choice of a particular drafting tool will depend on the intended outcome. In this way, I have chosen to use a touch-sensitive computerised screen because of the many advantages this has for my chosen outcome – fast, easily produced images which can be quickly copied and distributed. It is also possible to replicate most styles of drawing from a paintbrush to a 0.1mm ink pen. There are thousands of colours, opacities and textures available and work can be repeated, layered and filed quickly and efficiently. In short, the natural skill of the artist can be enhanced in many ways and not least by using the latest technology. There are disadvantages though. The technology itself is an inhibiting factor – the super-smooth screen surface leads to a lack of control. The formality of the process itself can inhibit the artist and reduce confidence in the outcome.

In the end, there is no substitute for the informality and spontaneity of a soft pencil on paper.

*

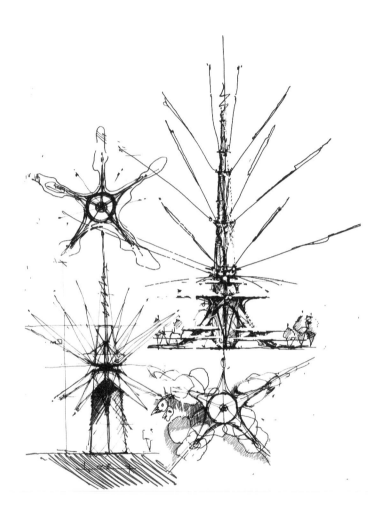

FIGURE 2.31
Urban Oasis
Source: Laurie Chetwood (Chetwood Associates).

FIGURE 2.32
Perfumed Garden, Chelsea
Source: Laurie Chetwood (Chetwood Associates).

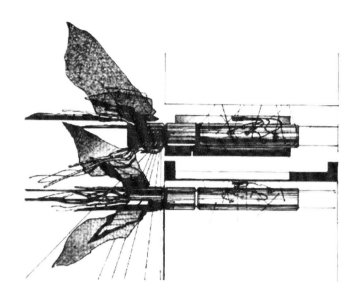

FIGURE 2.33
Studies for the Butterfly House
Source: Laurie Chetwood (Chetwood Associates).

FIGURE 2.34
Studies for the Butterfly House
Source: Laurie Chetwood (Chetwood Associates).

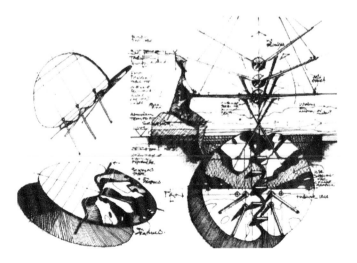

FIGURE 2.35
Studies for the Butterfly House
Source: Laurie Chetwood (Chetwood Associates).

FIGURE 2.36
Studies for the Butterfly House
Source: Laurie Chetwood (Chetwood Associates).

FIGURE 2.37

Studies for the Butterfly House

Source: Laurie Chetwood (Chetwood Associates).

2.5   Thinking on paper

Design and sketching

Successful dialogue between engineer and architect will be enhanced by an understanding of the design process and is further improved by sharing and developing ideas. This is where sketching comes into its own. Sketching almost forces a designer to crystallise ideas on paper, to think on paper, to explain and to clarify. A sketch can be quick and do the work of thousands of words. A series of sketches can be used to show how an idea was developed or a single sketch can be used to justify an adopted solution or strategy.

A perfect example is the early development of the Shard, a tower built in Southwark over London Bridge Station, the oldest major railway station in London. The first meeting between developer Irvine Sellar and architect Renzo Piano took place in Berlin on 30 May 2000. An idea of a building serving as a 'small vertical city' was conceived that day, a classic sketch on a napkin. It was taken back to the office and to London, where more sketches were produced making reference to a shape that would appear as if it was 'generated by the movement of the tracks on the ground', by the river of steel (the railway tracks) and by the river itself, the Thames. The project took shape by reference to the movement and scale of these 'rivers' and by inspiration drawn from images of soaring spires that once were a striking part of the ancient skyline of London.

Another technique used by some leading engineers requires far less artistic ability, and is based on experience, feel for structures, basic rules about height to width limits, span/depth ratios and knowledge of what is constructible. It is invaluable for early conceptual thinking and often involves taking a floor plan or an elevation or both from concept sketches and in rough freehand, scribbling on where shear walls, columns, cores or transfer structures are needed. The end result isn't always pretty but can soon be turned into something convincing.

Design and the built environment

Design in the world of construction is and always will be a challenge – it can be hard work, irritating and frustrating but more often than not it is fascinating, absorbing and interesting. Good design is creative and, given a fair wind and with the application of effort and dedication, is satisfying and rewarding.

The fundamental reason why it is such a challenge is because it has to cope with the practical issues of building in a complex world and must involve people from different backgrounds and disciplines.

It has long been recognised that the relationship between architect and structural or civil engineer (the engineer from now on) is one of the most important in the design of building or civil engineering projects. This is not surprising since the two disciplines have really developed from a common origin, the master builder. Experience shows that successful dialogue between architect and engineer is an essential part of producing good designs.

Good design in the practical world cannot be carried out in isolation. For instance, form finding using pure mathematics produces beautiful shapes but does not often meet the needs of practical buildings. Flat parallel and orthogonal planes of architecture based on the Cartesian box are usually easier to

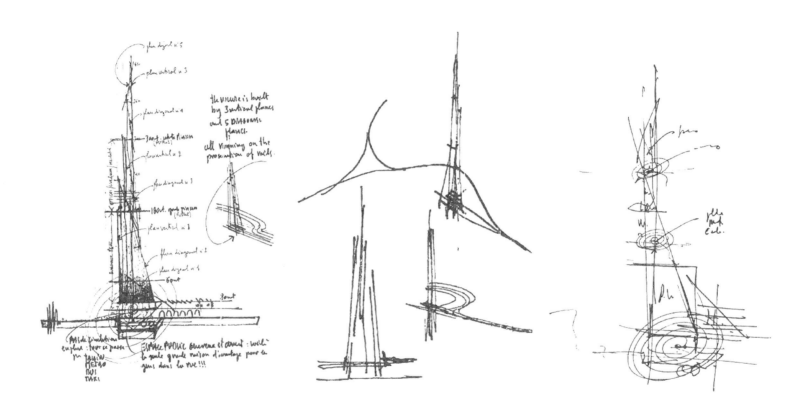

FIGURE 2.38

The Shard: scale and rivers of steel and water

Source: Renzo Piano

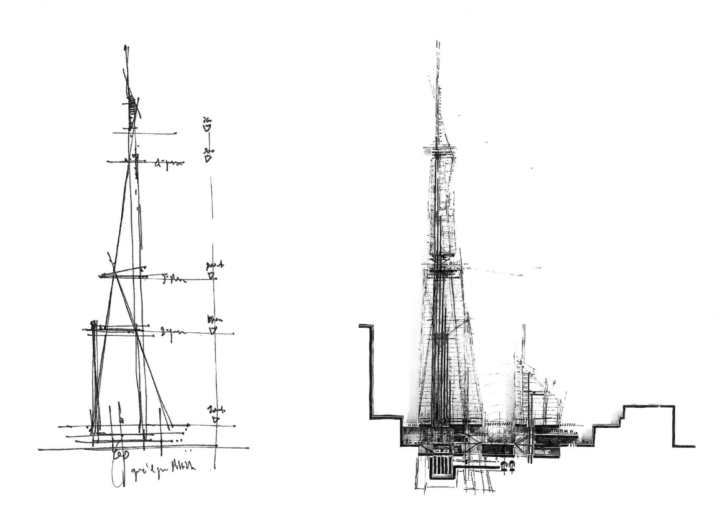

FIGURE 2.39
The napkin sketch and its derivative
Source: Renzo Piano

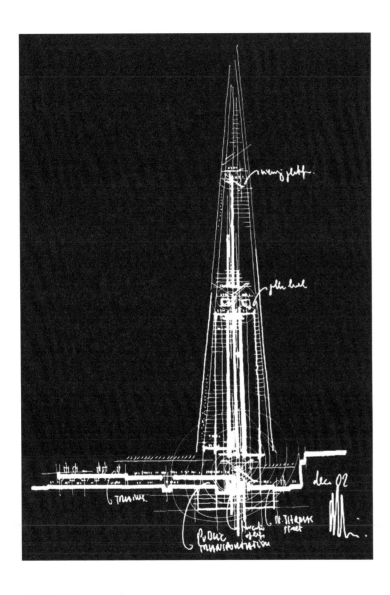

FIGURE 2.40
More evolution
Source: Renzo Piano

construct than architecture based on curved surfaces. It can be argued that mathematically generated surfaces with some unique geometric properties make them well suited for architectural use. There is a growing appreciation of the structural potential of doubly curved surfaces, exploited in the 1950s and 1960s by Candela and his thin concrete shells, for example. The underlying area of mathematics has advanced since the advent of computers and has significantly aided the study of minimal surfaces (surfaces with zero mean curvature; soap film type surfaces). However, their symmetry is often a problem. To be useful for architecture a geometric system needs a degree of flexibility, the ability to adapt to varied boundary conditions. It is no doubt useful to know about these surfaces (rheotomic surfaces – from the Greek *rheo*, flow, and *tomos*, cut or section, as in tomography), and instances where they have been used successfully, for example Nervi's 'lines of force' in the reinforcement pattern for the ceiling of his Gatti Wool Mill, but the opportunities to use them in everyday life are few and far between.

There are less esoteric pointers to good design. First, there is the old adage that form should follow function. It is difficult to deny that a steam locomotive often looks good but still there are superb looking machines and rather ugly ones. Another interesting analogy is the theory surrounding the Amesbury Archer. His skeleton was found near Stonehenge and he had clearly suffered a severe injury to his left knee; but to compensate, his right femur had grown in size and strength; strength where needed – a clear but less appealing example of form following function.

So sketching out how something functions may lead to efficient or aesthetically pleasing structural or building forms. This approach is closely allied to another useful strategy where reference is made to mathematical analysis rather than relying on geometry or function. Here shapes and sizes are chosen to respond to stresses and strains predicted in the structural analysis models. Deep members are used where bending moments are high, compression members are designed to prevent buckling, and thin strong members are used where parts of the structure are in tension only.

Yet another strategy is to mimic or 'reverse engineer' naturally occurring shapes and sizes where nature has already provided efficient and beautiful designs. In *On Growth and Form*, D'Arcy Wentworth Thompson refers to the example of Culmann's crane, a form inspired by the curves in a section of trabecular bone. Also, Mitchell trusses – certain 'ideal' structures arising through a process of topological optimisation – exhibit a characteristic arrangement of curved lines.

Physical models can also be used to explore ideas, but scale effect is a problem that has to be borne in mind – scaled up origami does not work. Linear relationships between size, structural behaviour and applied loadings do not exist. For example, corrections applied to measurements of a model in a wind tunnel to determine corresponding values for a full-sized object must be made using the science of fluid mechanics.

Galileo was aware of non-linearity of scaling and described the square/cube law – while area increases with the square of a dimension, the volume – and hence the mass – increases as the cube of the scale. He used this observation to explain that the limited strength of endoskeletons and exoskeletal material means the size of animals cannot be scaled up indefinitely. Two-metre long insects are not possible, at least on this planet.

There is no need to be over-sophisticated for day to day design, but a working knowledge of these fundamentals of structural engineering is important and of course should include a basic understanding of bending moments, shear forces, the interaction of struts and ties and load paths. An appreciation of how tendons, ligaments and spinal columns work, that hollow bones are strong and light, and that joints (connections) between structural members are difficult to design but critical, is an advantage and can add to the meaningful exchange of ideas between designers from different disciplines.

Ideas, structural form and materials

It is possible for designers from different disciplines to work together using only sketches, even if they do not understand each other's technical language. However, it does of course help to know the vocabulary of each discipline involved. As far as structural engineering is concerned, terms such as 'frame', 'arch', 'beam', 'dome', 'plate', 'membrane' and 'cantilever' are part of everyday speech and need no further explanation. Even a few additions, for instance about structural form, help a great deal. (See Figure 2.41.)

Structural Form	Diagram
Cable stay(ed) – flat structure in bending supported from above by cables radiating downward from masts that rise above.	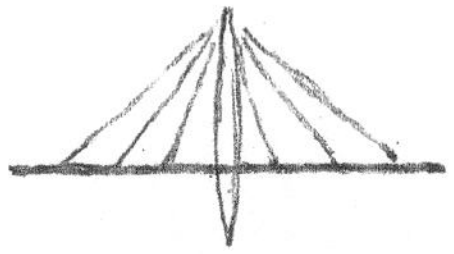
Catenary – a curve formed by a wire, rope, or chain hanging freely from two points that are not in the same vertical line.	
Funicular – a polygonal form finding structure either in pure compression or pure tension, supporting or carrying load at defined node points.	
Gridshell – a structure which derives its strength from its double curvature (in the same way that a fabric structure derives strength from double curvature), but is constructed of a grid or lattice. The grid can be made of any material, but is most often wood or steel.	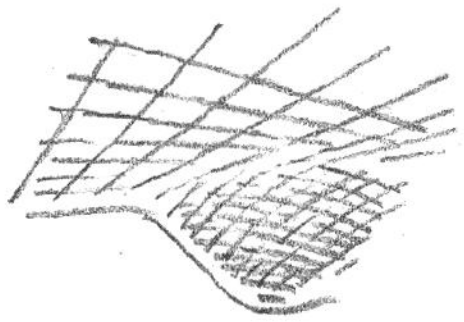
Diagrid – a supporting framework in a building formed with diagonally intersecting ribs.	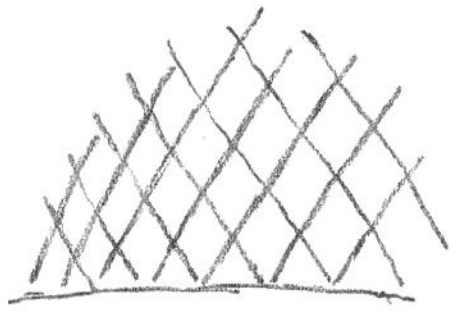
Shell – a thin shell is defined as a shell with a thickness which is small compared to its other dimensions and in which deformations are not large compared to thickness.	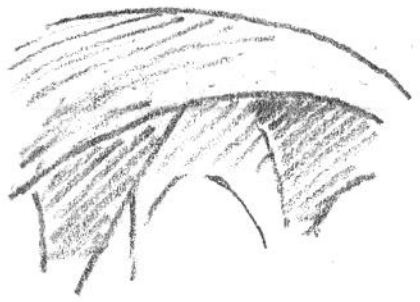

Structural Form	Diagram
Tensegrity – the characteristic property of a stable three-dimensional structure consisting of members under tension that are contiguous and members under compression that are isolated.	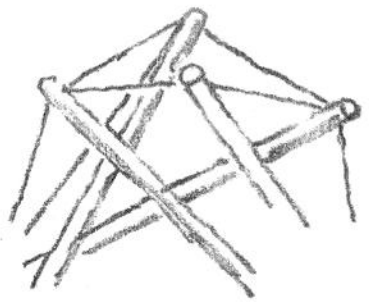
Monocoque – a structural approach that supports loads through an object's external skin, similar to a ping-pong ball or egg shell.	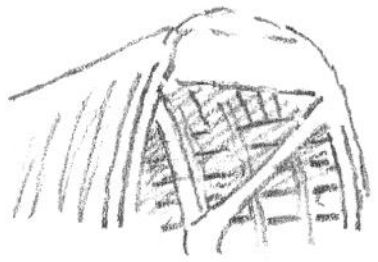
Folded plate – a thin-walled building structure of the shell type. Folded plate structures consist of flat components, or plates, that are interconnected at some dihedral angle (the angle created by two intersecting planes).	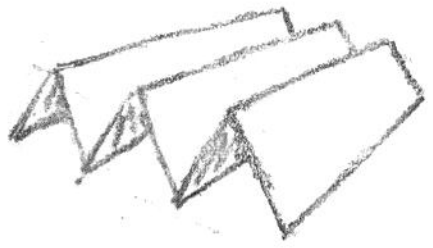
Geodesic – a geodesic dome is a spherical or partial-spherical shell structure or lattice shell based on a network of great circles (geodesics) on the surface of a sphere.	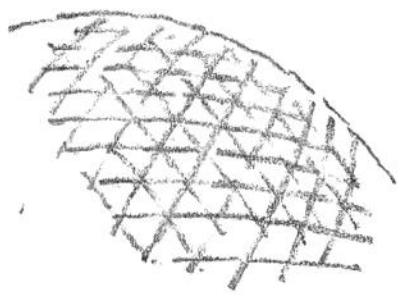
Anticlastic – having opposite curvatures at a given point; specifically: curved convexly along a longitudinal plane section and concavely along the perpendicular section – used of a surface – opposed to synclastic.	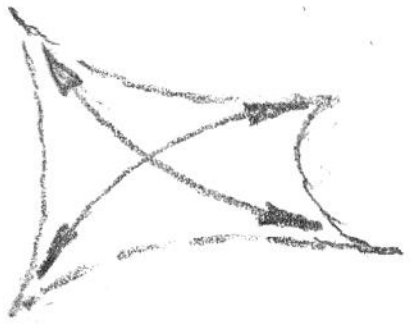
Vierendeel – an open-web truss with vertical members but without diagonals and with rigid joints.	

FIGURE 2.41
Structural form

Lightweight versus 'normal' structures

Another differentiator worth bearing in mind when sketching ideas, is whether a structure will be designed by regular codified rules and fits into a 'normal' category or whether a different approach has to be used. Small, lightweight structures are far from normal – simple use of wind, lateral and gravity loads isn't sufficient. Design of small, lightweight structures is often about notional loads, acceptable movement, dynamics, redundancy and above all, safety. Pure engineering judgement plays a bigger part, but nevertheless the logic behind a design should be described and committed to paper.

A good example is the 'Urban Oasis' design for Chetwood Architects (q.v.). If the arms or petals had been designed for normal wind loads, even a low pressure based on a 1 in 1 year wind speed, the supporting structure would appear heavy and grossly over-designed. The simple answer here was an arrangement whereby the arms were folded up against the mast if strong winds were forecast, out of harm's way. Other strategies for designing 'small and light' structures are to make use of prototyping, testing and refinement.

*

2.6   The use of hand sketches in practice

Developing ideas and concepts

In his *Second Sketchbook*, Tony Hunt says he does not feel comfortable thinking about a design problem without a pencil in his hand. For engineers and architects, sketching has long since been an essential part of the design process and problem solving. It is a recognised way of exploring and developing ideas that start in the head but go through the hand on a lengthy journey of exploration, explanation and development before they are embodied in reality.

It is not surprising that in the early stages of this process, the sketches are often no more than scribbles, many of which are discarded and committed to the waste bin. Design is iterative and very rarely a straightforward linear process.

Sketching is an effective way of explaining spatial relationships, visualising complex analysis, showing how ideas have evolved and demonstrating how something is going to be built.

From a practical point of view, there is no better way to start than finding a tracing roll and a soft pencil and downloading ideas onto paper in a rough and ready fashion, just to develop your own thoughts and then help begin a visual dialogue with colleagues and designers in related disciplines.

Design references

Structural form is not often derived from a mathematical analysis of a shape or a surface; neither is it common for it to spring from, for example, knowledge of biological systems found in nature. More often than not, spatial and functional requirements take centre stage, and reference to other influences is used to develop initial ideas and move the design in a particular direction. But in a way, that is the nature of design; you come up with a solution and check to see whether it is suitable or not by looking back and ensuring it meets the initial brief. Shape and design of structural members is often more closely related to reference material. The depth of beams for instance is related to loading and span, in the same way that the diameter of a branch in a tree depends on how strong the wood is, the weight that has to be carried and the span of the branch acting as a cantilever.

Design references are nevertheless important; sometimes they can move a concept forward, sometimes they can be used as a comparator, and sometimes they are valuable as retrospective rationalisation. Many concept sketches have been drawn by very famous designers after buildings have been built.

Design development

Designs do change for very good reason, for instance when other disciplines join the team with their own particular requirements or when, say, contractors look at buildability and how a structure or building can be put together in a safe and economic way. In fact, change often happens as you draw, as you develop ideas and test them on paper. In a complex world, don't expect to get things right first time!

Recording and explaining this type of design development can help show the design process has been thorough and rigorous. Saving the sketches which go along with this sort of design development can show, for instance, that safety has been given due consideration. They will act as an audit trail through an often very complex and convoluted process.

I had a manager once whose mantra was 'get it right first time'. I knew from that point, that he had crossed the line from being a designer to being a manager interested primarily in efficiency and economics.

Illustration and clarity

In engineering sketching, clarity is vitally important. Single ideas can be separated out, unnecessary detail can be stripped away and line weight and quality can be chosen to lay emphasis on the key ideas that you are trying to portray. There are a number of ways of achieving clarity which in turn deal with complexity:

- Assembly drawings: showing how elements fit together, rather like the diagrams beloved by model makers or flat-pack furniture manufacturers;
- Sequence sketches: step by step illustrations of how something is built against a time line;
- Cutaway drawings: the best examples and most influential as far as I was concerned were the double page drawings in the *Eagle* – a comic long since absent from the newsagents.

Beyond sketching?

In my experience you can't beat physical models; somehow they are still superior to the very best computer generated images; they are tangible. As a digression, physical models can also be used to test structural behaviour as well as how something is put together. Bill Addis's fascinating paper 'Toys That Save Millions: A History of Using Physical Models in Structural Design', published in the *Structural Engineer* in April 2013, describes the benefits but as he points out, it is not correct to assume that the behaviour of a model test can be scaled up linearly to full size. There are two types of structural phenomena or behaviour –

The CAERPHILLY CASTLE

IN 1923, under the supervision of the Chief Mechanical Engineer of the Great Western Railway, Mr C. B. Collett, there emerged from Swindon a locomotive that laid claim to being the most powerful passenger engine in Britain – No. 4073, *Caerphilly Castle*, the first of the famous 'Castle'-class locomotives which were to be the backbone of the motive power for passenger trains of the G.W.R. for many years.

Developed from the 'Star'-class engines in which the previous Chief Mechanical Engineer, Mr G. J. Churchward, had embodied such revolutionary ideas as the high-pressure taper boiler, and four cylinders with long-travel piston valves, the 'Castles' proved themselves worthy of the tasks to which they were put. They handled heavy express trains to the West of England, Wales and the Midlands, including the famous *Cheltenham Flyer*, for some time the world's fastest train.

Caerphilly Castle was exhibited at the Empire Exhibition at Wembley, in 1924, and now this magnificent locomotive, which proudly carried the crest of the Great Western Railway for so many years, has been preserved in the Science Museum in London, where it will form part of the improved Rail Transport Collection which will be opened to the public later on.

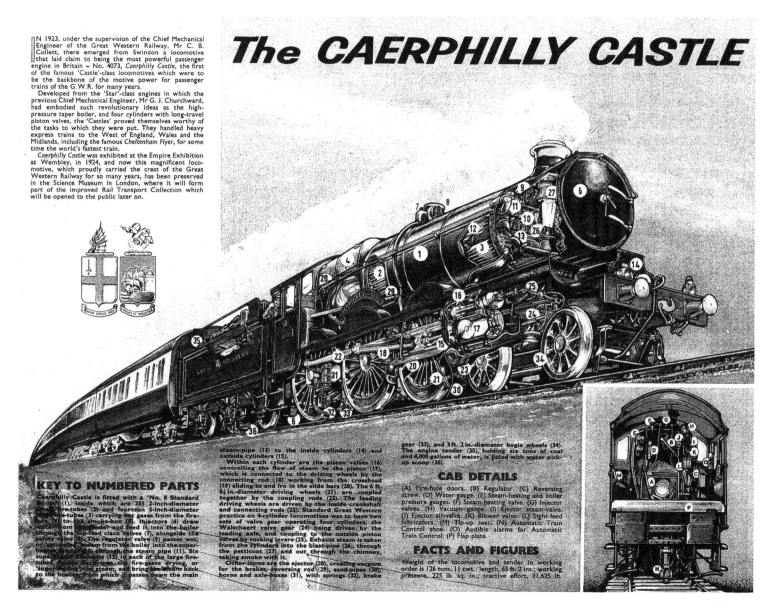

KEY TO NUMBERED PARTS

Caerphilly Castle is fitted with a 'No. 8 Standard' boiler (1), inside which are 201 2-inch-diameter small fire-tubes and fourteen 5-inch-diameter large fire-tubes (3) carrying hot gases from the fire-box (2) to the smoke-box (5). Injectors (4) draw water from the tender and feed it into the boiler through the top-feed clack valves (7), alongside the safety valve (8). The regulator valve (9) passes wet, or saturated steam from the boiler into the super-heater header (10) whence the steam-pipe (11). Six superheater elements (12) in each of the large fire-tubes carry this back into the fire-gases, drying, or superheating, the steam, and bring the steam back to the header from which it passes down the main

steam-pipe (13) to the inside cylinders (14) and outside cylinders (15).

Within each cylinder are the piston valves (16) controlling the flow of steam to the piston (17), which is connected to the driving wheels by the connecting rod (18) working from the crosshead (19) sliding to and fro in the slide bars (20). The 6 ft. 8½ in.-diameter driving wheels (21) are coupled together by the coupling rods (22). The leading driving wheels are driven by the inside crankshaft and connecting rods (23). Standard Great Western practice on 4-cylinder locomotives was to have two sets of valve gear operating four cylinders; the Walschaert valve gear (24) being driven by the leading axle, and coupling to the outside piston valves by rocking levers (25). Exhaust steam is taken from the cylinders into the blast-pipe (26), through the petticoat (27) and out through the chimney, taking smoke with it.

Other items are the ejector (28), creating vacuum for the brakes, reversing rod (29), sand-pipes (30), horns and axle-boxes (31), with springs (32), brake

gear (33), and 3 ft. 2 in.-diameter bogie wheels (34). The engine tender (35), holding six tons of coal and 4,000 gallons of water, is fitted with water pick-up scoop (36).

CAB DETAILS

(A) Fire-hole doors. (B) Regulator. (C) Reversing screw. (D) Water gauge. (E) Steam-heating and boiler pressure gauges. (F) Steam-heating valve. (G) Injector valves. (H) Vacuum gauge. (I) Ejector steam-valve. (J) Ejector air-valve. (K) Blower valve. (L) Sight-feed lubricators. (M) Tip-up seat. (N) Automatic Train Control shoe. (O) Audible alarms for Automatic Train Control. (P) Flap plate.

FACTS AND FIGURES

Weight of the locomotive and tender in working order is 126 tons, 11 cwt.; length, 65 ft. 2 ins.; working pressure, 225 lb. sq. in.; tractive effort, 31,625 lb.

FIGURE 2.42

Hand-drawn *Eagle* cutaway

those that can be scaled up linearly, such as the linear dimensions of a structure, the shape of a hanging chain of weights or a membrane, and of funicular arches, vaults and domes; and the stability of masonry compression structures, including arches, vaults and domes – and those that cannot, such as the mass of a structure, the strength and stiffness of a beam, or the buckling load of a column or thin shell.

Addis says:

> This observation explains why masonry structures were able to develop so spectacularly, long before any scientific or mathematical understanding of structural behaviour. A model arch, vault or dome can be reliably used to predict the behaviour of a similar, full-sized structure.

> Galileo mentions the non-linearity of scaling (the square/cube law): area increases with the square of the scale factor, the volume and hence the mass, increases as the cube of the scale. He used this observation to explain that the limited strength of bone material means the size of animals cannot be scaled up indefinitely.

There are mathematical techniques based on dimensionless numbers that can be used, but it is not possible to make a single scale-model structure that represents the full-size structure and its behaviour in every respect.

Coming back to sketching and drawing, most physical models start off with a sketch and then a drawing.

- Exploded diagrams: show an assembly taken apart;
- Separating ideas: perhaps using copies of the same sketch to show different aspects of the design (load paths, transfers, areas of overlapping structure, loadings, areas of high stress, access routes, air paths and so on);
- Storyboards and cartoons: just a visual way of explaining a design or a strategy or a thought process.

The techniques can be used equally well for new-build projects and for understanding how existing buildings are put together and how they work.

*

PART 3: SKETCHBOOK

3.0 INTRODUCTION

The following pages show examples of how hand sketches are put to various construction-related uses – they are taken from work on bid submissions and from various engineering/architectural design stages ranging from concept to construction and beyond and were not drawn specifically for this book. Most of the sketches are constructed using simple axonometric projection and all were produced to illustrate engineering or combined engineering and architectural principles or objectives. All aim to show a personal touch and a level of human involvement which avoids some of the pitfalls of overly clinical, unfeeling or over-complicated CAD imagery. If the sketches have anything in common, it is that they are attempts at putting ideas on paper with as much clarity as possible. They are intended to communicate, to assist in sharing and exploring thoughts and ideas.

Few if any of the sketches are from design workshops; I find these are usually scribbles that are the first manifestation of ideas and too abstract to be of much value in themselves. The best approach is to take these scribbles away, and draw cartoons that trace the emergence of a design or approach. Examples are shown on pages 56, 60, 61, 67, 98, 106, 127, 139, 163, 190, 205, 241 and 242. Perhaps it is retrospective justification or rationalisation but it is design development, the way you and your team have decided to go, and it is good and reassuring to show you have followed a train of thought.

Some of the sketches were produced to summarise a proposal at a particular stage of the design development. They are again useful, even as an audit trail, to show how a concept, often necessarily simplistic, has been retained, developed or, with good reason, has been lost altogether.

The sketches that make up Figure 3.0 tell a story. The left hand drawing was produced in 1970 for a contractor working on the redevelopment of Fort Regent in Jersey who was rightly very proud of his newly acquired demolition machinery. I think it was then that I realised that civil engineering plant are great fun to draw. The sketch on the right, from forty-five years later, is part of a design development note still based on hand sketching, but this time produced for a major scheme to reinvigorate the leisure facilities at the Fort.

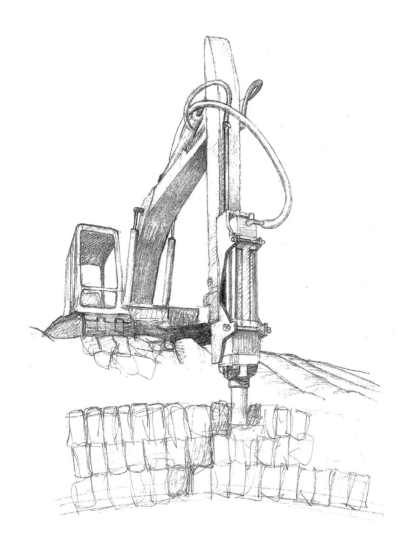

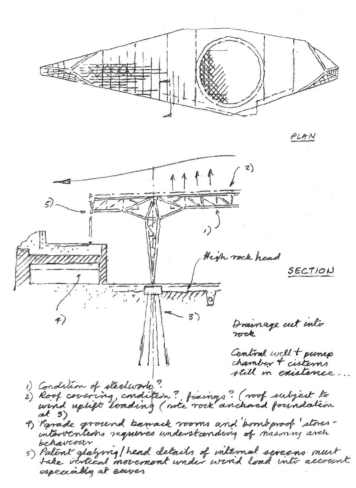

PLAN

SECTION

High rock head

Drainage cut into rock

Central well + pump chamber + cisterns still in existence...

1) Condition of steelwork?
2) Roof covering, condition?, fixings? (roof subject to wind uplift loading (note rock anchored foundation at 3)
4) Parade ground barrack rooms and 'bombproof' stores- interventions requires understanding of masonry arch behaviour
5) Patent glazing/head details of internal screens must take vertical movement under wind load into account especially at eaves

FIGURE 3.0
Fort Regent, Jersey

Building types

Because the sketches were made primarily for proposals, competitive bids or in connection with the various work stages of real projects, they were often drawn with more than a single objective in mind. The sketches are therefore best grouped according to building type rather than for specific purpose, and each is provided with a description of why they were produced in the first place. Note the number of sketches that illustrate basement construction in urban areas. This is indicative of just how difficult works in the ground can be in terms of engineering design, site logistics, economics and buildability.

*

3.1   LONG SPAN STRUCTURES

The development of these buildings started in the nineteenth century with the advent of the railways, which generated the need for long span enclosures at a time when the basis of building technology was advancing from the use of cast iron to wrought iron and then to steel. Steel, with its high strength to weight ratio, was and still is the ideal material for the construction of long span lightweight structures. Train sheds in the nineteenth century grew from Brunel's Paddington Station with a main span of 30m to Barlow and Ordish's St Pancras Station with a span of 74m. Exhibition halls followed suit, from Paxton's Great Exhibition hall in 1851 (later re-erected at Crystal Palace) to the Galerie des Machines for the 1889 Paris Exhibition by Contamin and Dutert with a span of 114m.

In the early twentieth century the major technological advances changed from land to air, starting with the development of the airship. The need to house these huge machines led to the construction of economical, long span, large volume sheds – the largest was the Goodyear Airdock at Akron, Ohio, USA, opened in 1929, with a parabolic three-pinned arch structure spanning 99m.

Larger heavier than air aircraft have also forced the development of clear-span hangars – when our practice worked on a scheme for a hangar for the first 747s, we christened it a 'glove hangar' because it was designed with a special housing for the tail fin.

So when is long span long span? It depends on function but warehouses, train sheds, sports stadia and even supermarkets have all qualified in the past. The structural form tends to be varied in buildings; less so in bridges – long span building structures often span in two directions while bridges on the whole are linear. Common types of long span building structures are listed below:

- Various beam types – single span and continuous spans
- Portals
- Arches
- Domes
- Masted structures
- Two-way spanning structures
- Space decks
- Cable nets
- Membranes
- Umbrellas
- Shells
- Span shortening back-stayed cantilevers
- Air supported structures.

Although steel is by far the most widely used material, long spans have been built using concrete, steel/timber composites, timber, and PTFE. Achieving a cost effective design for a large span roof often involves lightweight structures, making these roofs highly sensitive to wind and snow loading. In order for sophisticated 3D structural modelling to deliver efficient structural design, accurate loading scenarios are required for wind that account for complex fluid–structure interactions including wind driven dynamic effects.

In wind dominated long span designs, wind load reversal, asymmetrical or patch loading from snow or wind, needs to be considered carefully and so does the effect of inevitably large deflections on, for example, suspended services or hangar doors. Wind related environmental effects such as pitch microclimate and spectator comfort in stadia design and external microclimate around large buildings all require careful consideration from an early stage.

FIGURE 3.1/1

NATO headquarters building. A competition sketch for a long span, lightweight rainscreen roof over a set of free standing buildings for NATO. The concept, never fully supported by serious analysis and calculation, was based on perimeter 'span shortening' cantilevers. They supported an inner tension ring beam which in turn supported a geodesic dome with its central compression ring. Asymmetric wind and snow loading would have been difficult to deal with and constructing the conventional buildings under the roof would have led to craneage difficulties.

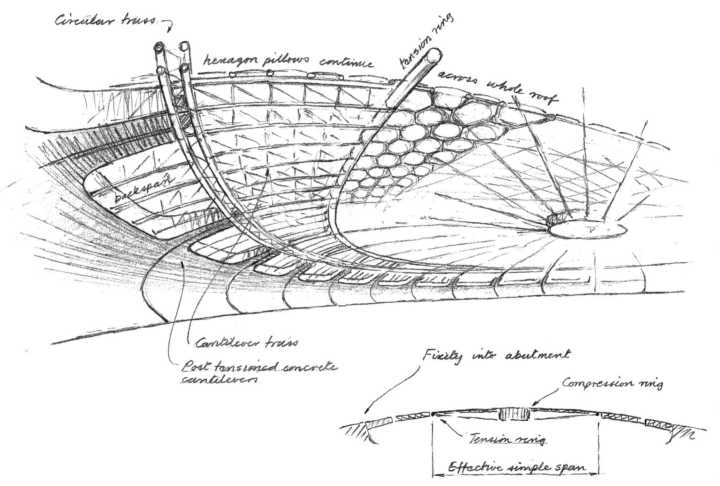

Circular truss

hexagon pillows continue

tension ring

across whole roof

backspan

Cantilever truss

Post tensioned concrete cantilevers

Fixity into abutment

Compression ring

Tension ring

Effective simple span

The Science Museum, Wroughton, Swindon. Here the design progressed from an option based on a clear span arch to a number of options using propped or cable assisted spans. This approach was viable since the layout of the museum could easily accommodate internal supports without compromising exhibit or storage space. The design moved away from heroic structure towards smaller spans, which in turn led to much improved structural efficiency and opened up design possibilities with regard to supporting areas of green roof, sub-division of the structure into transportable elements and a visually interesting structural arrangement. Reduced spans with smaller span/depth ratios were also more suitable for the use of hybrid timber/steel trusses.

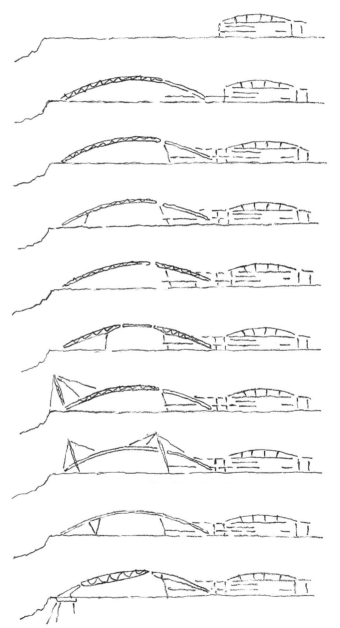

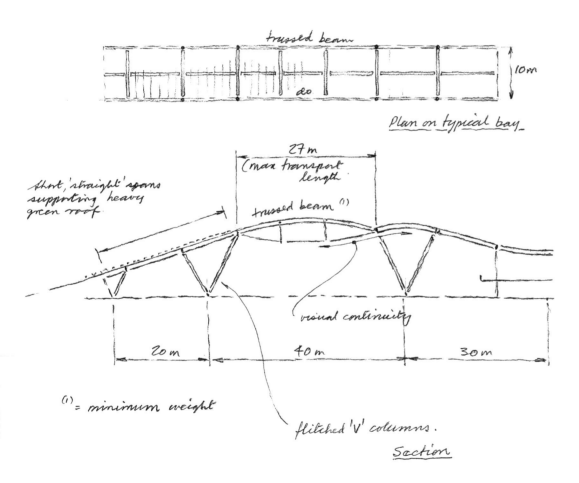

trussed beam

10m

do

Plan on typical bay

27 m (max transport length

short, 'straight' spans supporting heavy green roof.

trussed beam (1)

visual continuity

20 m   40 m   30 m

(1) = minimum weight

flitched 'V' columns.

Section

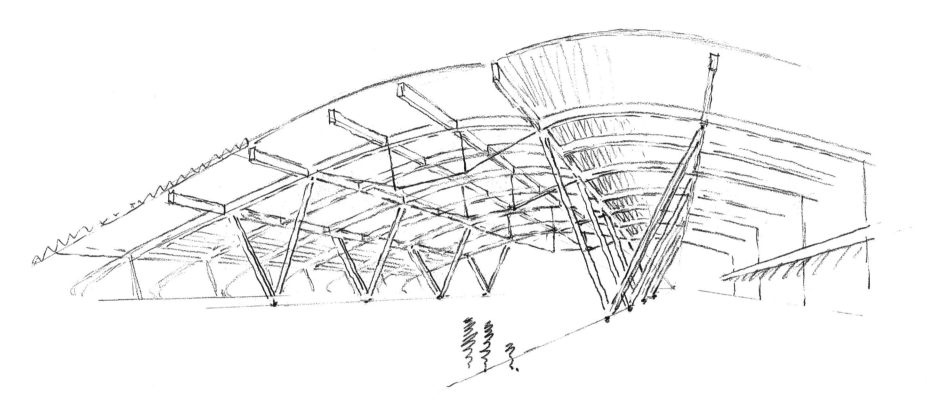

FIGURE 3.1/3

The Science Museum, Wroughton, Swindon. Funding was never achieved and the scheme was abandoned. This was a great shame because the preferred option in steel and timber would have provided an attractive structure and a somewhat unusual solution in the UK which has no great tradition in long span timber buildings. Further design development might have led to the introduction of natural light and an optimised structure using steel to carry tensile loads and timber to deal with the compressive loads. The whole approach could have produced a building with a highly sustainable pedigree.

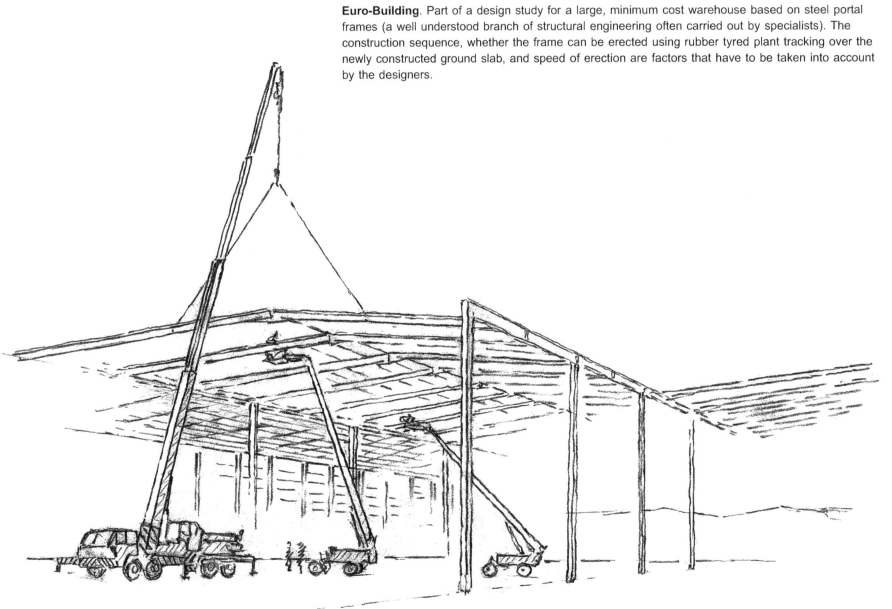

FIGURE 3.1/4

Euro-Building. Part of a design study for a large, minimum cost warehouse based on steel portal frames (a well understood branch of structural engineering often carried out by specialists). The construction sequence, whether the frame can be erected using rubber tyred plant tracking over the newly constructed ground slab, and speed of erection are factors that have to be taken into account by the designers.

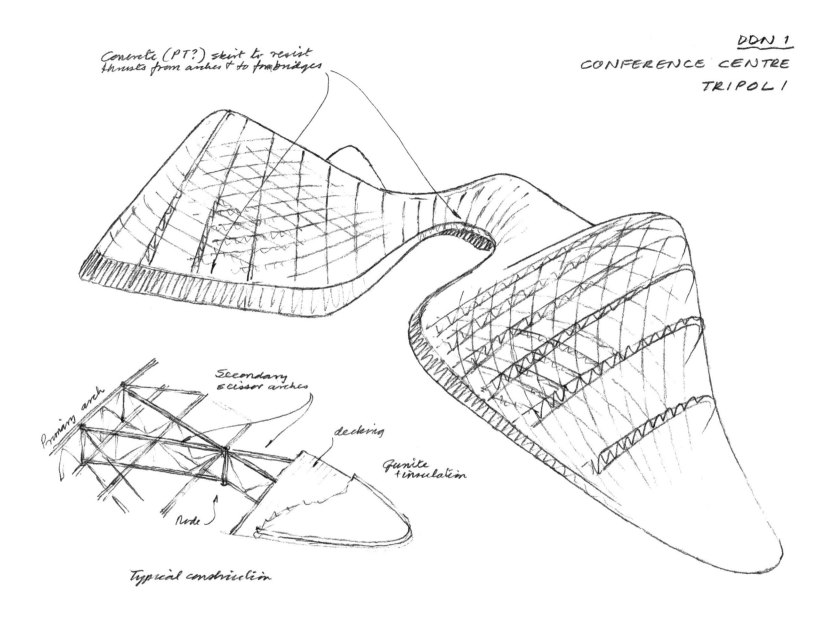

Concrete (PT?) skirt to resist thrusts from arches & to framebridges

Secondary scissor arches

Primary arch

decking

Gunite + insulation

Node

Typical construction

FIGURE 3.1/5
Conference centre, Tripoli, Libya. A proposal for a conference centre which was planned for a prominent site in Tripoli, Libya, by Zaha Hadid Architects. This grand design was designed to provide several venues all under the same roof. In its early stages of design development, the curvilinear roof structure was based on parallel primary trussed arches supported on inter-connecting secondary steel trusses forming a diagrid. Lateral thrusts were resisted by a concrete perimeter skirt tied from side to side by internal structure. The roof covering consisted of insulated gunite on deep profile metal decking.

FIGURE 3.1/6

Cambridge Regional College, phase 2. One of Powell & Moya's final schemes, this design developed from a grand gestured masted structure to a much more subtle one. Instead of a very extravagant central mast, paired masts were arranged between classroom blocks with tie-downs positioned either side of central circulation corridors. This has the effect of giving just a hint in the external appearance of what might be happening below the roof planes. The system gave spans that could be achieved using glulam beams over flexible workshop space, supported by cable assisted glulam cantilevers supported by the twin masts. The two dimensional concept was replicated longitudinally but the back-to-back structure is not inter-dependent, and therefore the building bays could be offset to suit site constraints.

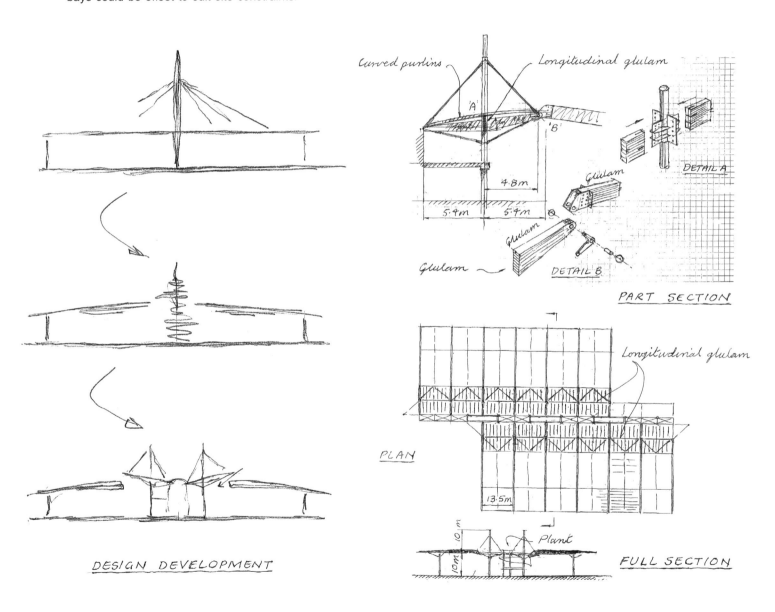

DESIGN DEVELOPMENT

Curved purlins

Longitudinal glulam

'A'
'B'

DETAIL A

4.8m

5.4m   5.4m

Glulam

Glulam

Glulam

DETAIL B

PART SECTION

Longitudinal glulam

PLAN

13.5m

Plant

10 m

10m

FULL SECTION

FIGURE 3.1/7

Exhibition centre. An exhibition centre in the Middle East, one of the largest of its kind with a 73,000 m² indoor event space. Its success is thought to have stimulated local development. This scheme was designed to deliver more indoor area by creating an arena between existing buildings. Site access and erection methodologies using mobile cranes were the main considerations in developing a fairly pragmatic design solution.

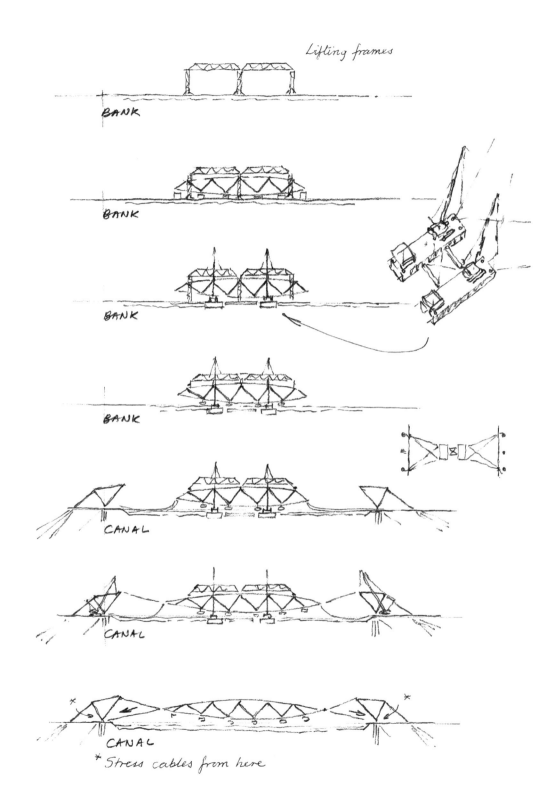

Lifting frames

BANK

BANK

BANK

BANK

CANAL

CANAL

CANAL

* Stress cables from here

FIGURE 3.1/8

Sustainable power generation, India. P4P developed an innovative scheme to construct photovoltaic arrays over irrigation canals in north-west India. The idea was based on lightweight, efficient cable structures, supporting PV panels over the canals, which would both reduce evaporation and minimise land take and thereby deliver a truly sustainable design. How to build the structures was the biggest challenge.

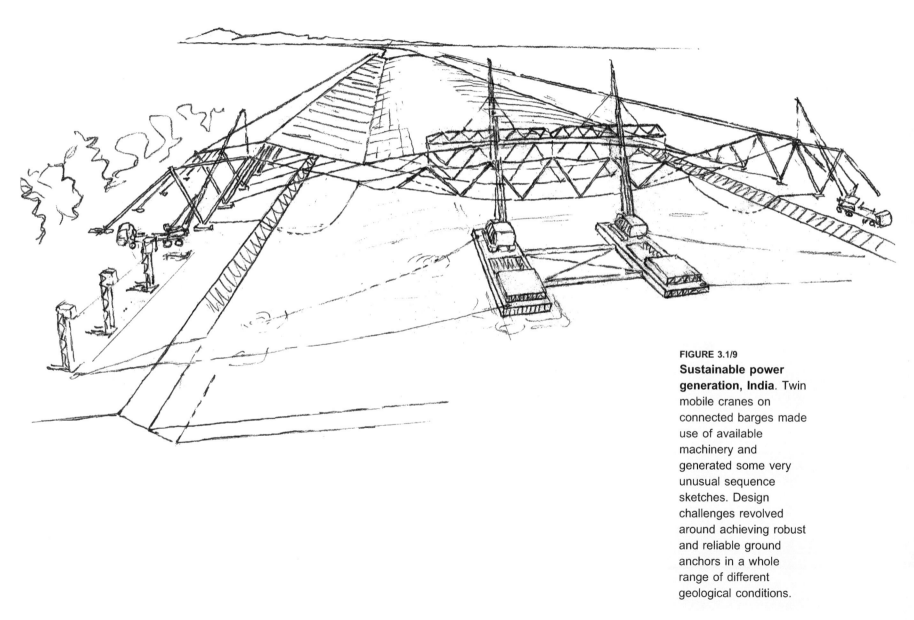

FIGURE 3.1/9

Sustainable power generation, India. Twin mobile cranes on connected barges made use of available machinery and generated some very unusual sequence sketches. Design challenges revolved around achieving robust and reliable ground anchors in a whole range of different geological conditions.

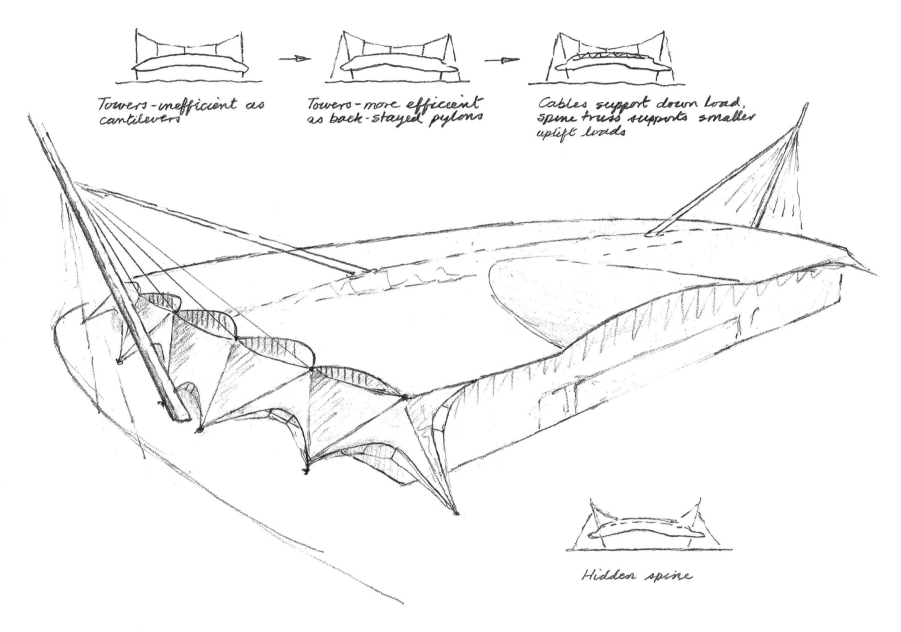

Towers - inefficient as cantilevers

Towers - more efficient as back-stayed pylons

Cables support down load, spine truss supports smaller uplift loads

Hidden spine

FIGURE 3.1/10

Retail store, Manchester. The primary aim of this scheme was to design a retail building with a column-free interior under a lightweight roof. The efficiency of the cable stay system relied on the way the masts were inclined, the backstay arrangement, the cable profile being not too shallow and whether the wind uplift loads were carried by counter-poised cables or wind trusses. Working with Chetwood Architects, the final proposal was based on a hidden wind truss and angled backstays.

FIGURE 3.1/11

J. Sainsbury, Camden Town, London. Early proposals for J. Sainsbury at Camden Town were considered too ordinary for the planners so Nicholas Grimshaw was commissioned to produce a more radical design. These sketches are a retrospective of white board sessions with the architectural team and show the development of ideas from two storey massing, positioning the upper level accommodation over the columns, portalising the structure, to an articulated system with back-stayed cantilever supports carrying an internal lightweight roof. Sketch reproduced by kind permission of Sainsbury's Supermarkets Ltd.

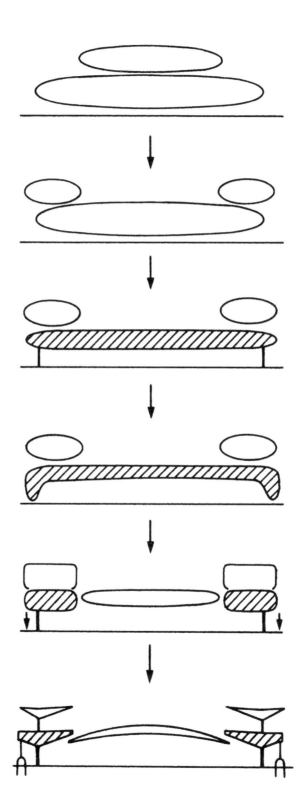

3.2   HIGH RISE

In recent decades, the planet has urbanised on an unprecedented scale. We need to create cities that can cope with these demographic changes and accommodate the inevitable pressure on transport, energy, water and living spaces. In an age when more than half the world's population lives in cities, there is more demand than ever for space. Tall buildings offer the most efficient use of land; building up, rather than out, must be part of the solution. It is part of the responsibility of engineers to design the structure of tall buildings so that they meet the various needs of the urban population in a flexible and responsible way. Consequently, the challenges for the engineer are numerous.

One of the key objectives of building tall in historic cities is keeping streetscapes permeable and delicate. Tall buildings require large cores and huge columns to take lateral and gravity forces, but they also must respect people and public spaces at street level. There is a history in architectural development of not paying sufficient attention to the interface of tall buildings and life at ground level. Now it is generally accepted by the best designers that the effect at street level is equally as important as the impact on the skyline, and is critical to integration into the culture and lifeblood of a city.

In the past, street design has not been a priority for all stakeholders – the net result is that tall buildings have been planted on sites without due regard for their surroundings. By engaging in a different approach, through an evolving and collaborative process with professional advisors, designers and more enlightened key stakeholders, a robust and respectful result can be achieved. Analysing different street typologies and public

realm treatments and assessing capacity for pedestrians, public transport and traffic now contribute to balanced streetscapes, in terms of movement and the psychological perception of tall buildings.

The concept of the 'vertical city' is gaining acceptance, resulting in more mixed-use tall buildings in recent years. Renzo Piano referred to the Shard as a vertical city. He recognised that tall buildings need to deliver places that people enjoy for living, leisure and shopping, as well as for working. The wider economic, social and environmental benefits of tall buildings set in good public realm is beyond question. A well-positioned tall building offering such benefits can have a regenerative effect on its surroundings.

So, apart from how a tall building fits into the grain and fabric of a city, and the scale of structure at street level, what else is different about the design of a true tall building? There are several issues that are particularly important for the structural engineer:

Design criteria

- Efficiency:
  - Viability usually depends on efficiency, which can in part be measured by calculating the building's 'net to gross' area – in other words, efficiency depends on the ratio of useable floor area to the gross floor area which includes the core (vertical transportation and risers), space for plant and so on. In a tall building the space taken for vertical

transportation and risers increases with height, as does the influence of the lateral system needed to support horizontal loads.

- Movement:
  - Horizontal movement under lateral loads (sway). The overturning effect increases as sway increases. Excessive sway can also lead to damage of façades and internal finishes.
  - Acceleration (this is what you feel when a tall building oscillates).
  - Differential axial movement in vertical members due to the use of different materials (say a concrete core and a steel frame) or due to different strains caused by non-uniform stresses in separate members carrying gravity loads.
- Gravity and lateral loads:
  - The sheer scale of structure needed to safely transmit gravity loads to the foundation and to resist wind or seismic loads must be taken into account.
- Floor-to-floor height – saving a few millimetres on every storey could mean an additional floor becomes possible within a given stack (overall height).
- Robustness – protection against disproportionate collapse due to localised damage.
- Protection against the effects of an extreme event.

Countless other design factors have to be dealt with, including (to name just a few):

- Aesthetics
- Impact on the cityscape
- Rights of light
- Fire strategy
- Lifting strategy
- Façade design
- Façade cleaning
- Plant space and location
- Integration of services
- Buildability
- Timescale
- Maintenance
- Cost.

In general, the structure plays a more important part in the conceptual design of towers than it perhaps does in other building types. An understanding of possible structural solutions, how a tall building is going to meet the ground and how it is going to be built is essential in the formative design stages of a project.

FIGURE 3.2/1

The Shard at London Bridge. Early research by the architectural team from the Renzo Piano Building Workshop included viewing the urban landscape from the top of Southwark Towers. RPBW's architects were struck by the view of a river of water (the Thames) and a river of steel (the railway tracks). The project then began to take shape, by reference to the movement and scale of these 'rivers', by the inspiration of seventeenth-century landscapes, and through a belief that the idea of a tall, mixed-use tower was an idea fully compatible with urban regeneration and living in the city.

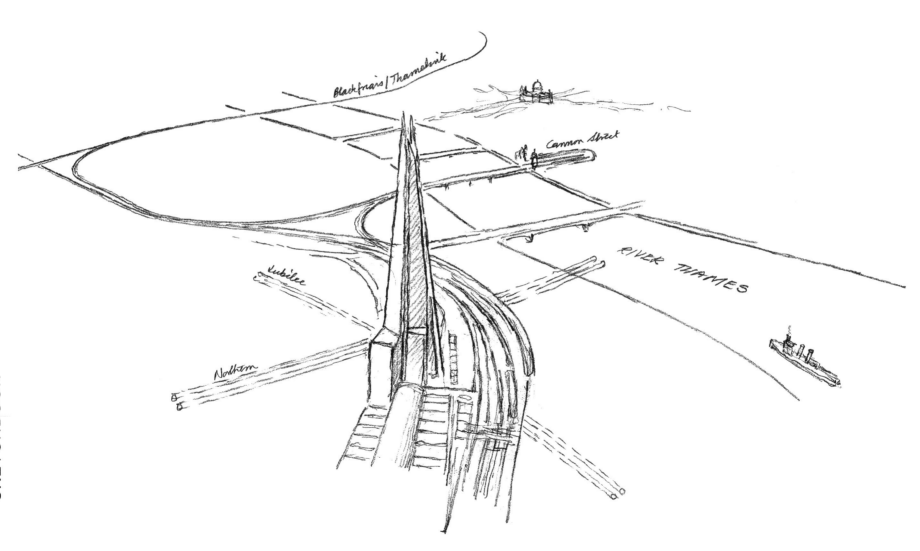

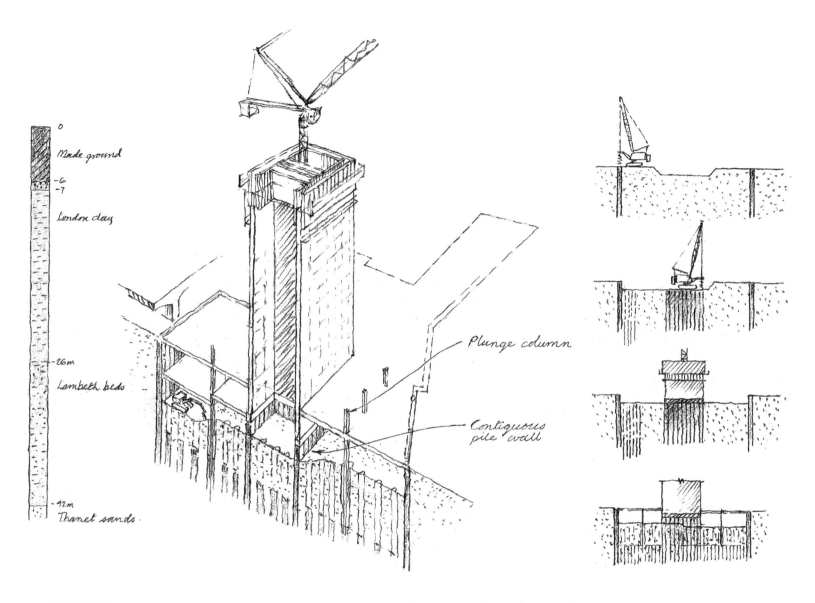

FIGURE 3.2/2

The Shard at London Bridge. Programme is always important and this sketch shows a 'top down' sequence where the main core and the substructure are being constructed simultaneously. First the perimeter basement wall is built followed by the core substructure. The ground level slab is then constructed and, with the walls now propped by the ground slab, excavation takes place below. This process is repeated until the lowest raft slab is reached and meanwhile, the core construction is underway from ground level upwards. Note that this proposal pre-dated any by contractors and their advisors who joined the team at later stages.

FIGURE 3.2/3

The Shard at London Bridge. Below ground constraints and boundary conditions were investigated in detail. The Jubilee Line tunnels run very close to the northwest corner of the site, vent shafts are close by and a disused lift shaft and associated tunnels crossed the site boundary at the southwest corner. On the northern and eastern boundaries the interface with the substructure arches of London Bridge Station had to be understood. In addition, piles from earlier buildings across the site and sensitive infrastructure in the road and pavements to the south all added to the complexity.

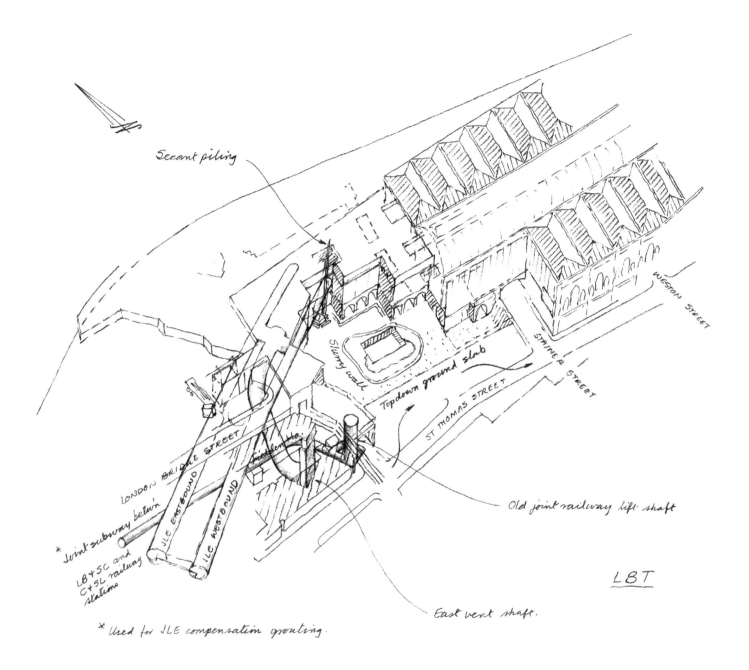

Secant piling

Slurry wall

Topdown ground slab

WESTON STREET

STAINER STREET

ST THOMAS STREET

LONDON BRIDGE STREET

JLE EASTBOUND

JLE WESTBOUND

* Joint subway betw'n

LB + SC and C + SL railway stations

Old joint railway lift shaft

East vent shaft.

LBT

* Used for JLE compensation grouting.

FIGURE 3.2/4

The Shard at London Bridge. From left to right, the earliest design for the lateral stability system was based on an all-steel framed building with a specially stiffened core. This evolved into a system that engaged the entire structure but with the need for a tuned mass damper at high level. Steel outriggers were added at plant level, connected to sizeable internal columns. Eventually, the upper part of the building was changed to concrete and following careful analysis of the distribution of mass, stiffness and damping within the building, all the normal design criteria could be achieved by the addition of a simple hat truss but with no need for a tuned mass damper at the top.

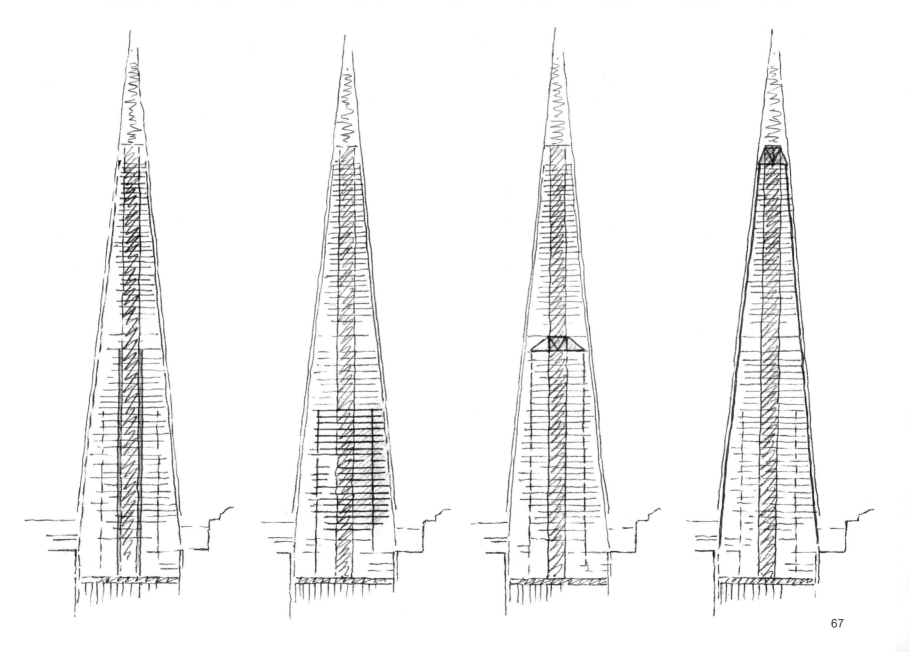

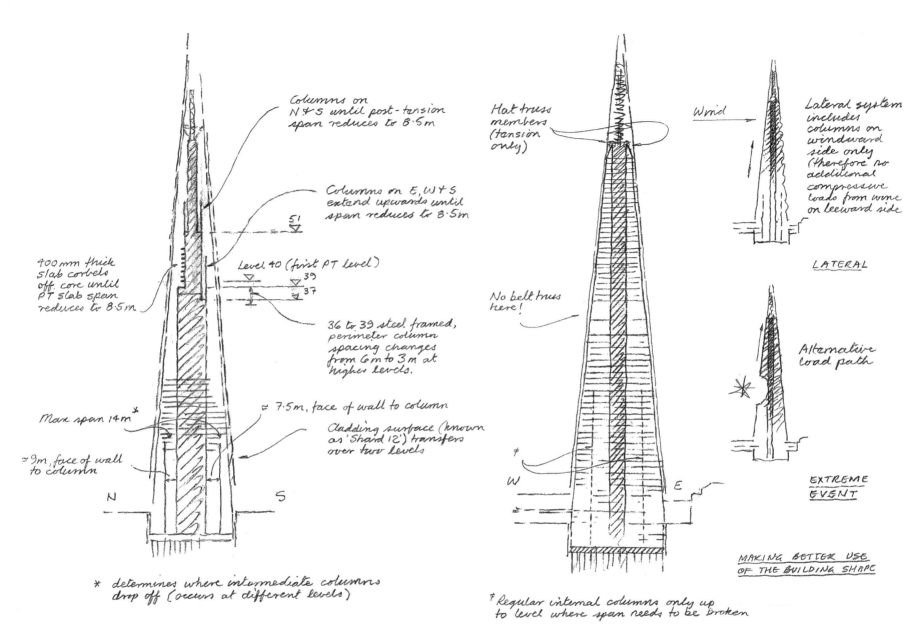

Columns on N & S until post-tension span reduces to 8.5m

Columns on E, W & S extend upwards until span reduces to 8.5m

400mm thick slab corbels off core until PT slab span reduces to 8.5m

51

Level 40 (first PT level)

39
37

36 to 39 steel framed, perimeter column spacing changes from 6m to 3m at higher levels.

≈ 7.5m, face of wall to column

Max span 14m *

≈ 9m, face of wall to column

Cladding surface (known as 'Shard 12') transfers over two levels

N   S

* determines where intermediate columns drop off (occurs at different levels)

Hat truss members (tension only)

No belt truss here!

W   E

‡ Regular internal columns only up to level where span needs to be broken

Wind

Lateral system includes columns on windward side only (therefore no additional compressive loads from wind on leeward side)

LATERAL

Alternative load path

EXTREME EVENT

MAKING BETTER USE OF THE BUILDING SHAPE

FIGURE 3.2/5
The Shard at London Bridge. A 'mixed structure' was developed for the 'mixed use' Shard, with steel in the lower (office) levels and concrete for the hotel and the residences at the top; and it made good sense to revert to steel for the spire. Many column arrangements were investigated, always with the aim of minimising the number of internal columns and, where possible, landing these columns on core walls at levels where lifts dropped off. Key features of the structural design were captured on this sketch.

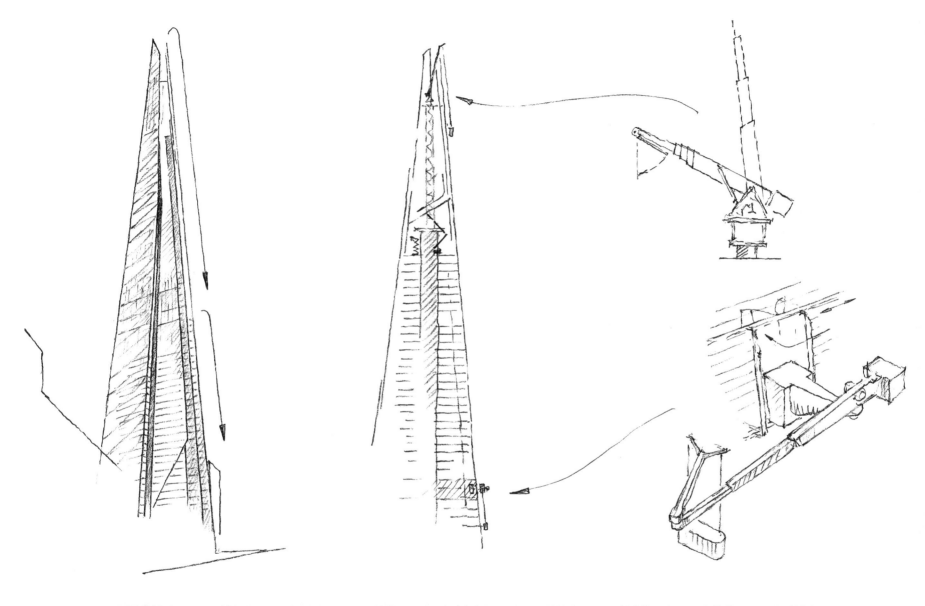

FIGURE 3.2/6

The Shard at London Bridge. Façade access is especially challenging on an asymmetrical tall building like the Shard. Vertical rope lengths for cleaning cradles are restricted by mandatory codes and therefore BMUs (Building Maintenance Units) must be located at more than one level, and are usually housed in areas reserved for plant rooms. The BMUs themselves are complicated pieces of machinery and have to be stowed away out of sight when not actually in use. As described later, the machine at the top of the spire was also designed to assist in the dismantling of the final construction tower crane.

The Shard at London Bridge. Like most tall buildings, the core of the Shard is multi-functional. It not only provides the key element of the lateral stability system but carries gravity loads, accommodates vertical circulation in the form of lifts and stairs and provides risers for mechanical and electrical services. Not surprisingly, a great deal of effort is required to find the most efficient arrangement; a compact core provides a good net-to-gross ratio for the floor plates. Quick and easy exploded axonometrics were used to explore the options and to study in particular set-back levels where lifts drop off and the core can reduce in size. Other key details were developed using axonometrics, including the spire, perimeter edge details and the form of the corner winter gardens. (The bottom right view shows the exposed structure in the underside of a winter garden.)

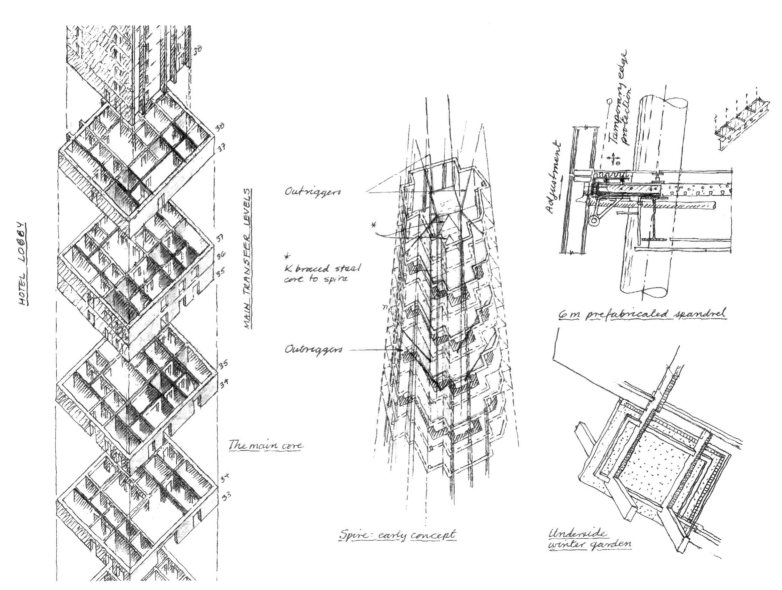

SKETCHBOOK

FIGURE 3.2/8

The Shard at London Bridge. The building footprint covers the entire site and the area is surrounded by one of the busiest railway stations in London, a bus station and busy, narrow streets. Craneage strategies were developed with the pre-construction advisor, Brookfield Multiplex, and due consideration was given to procedures where cranes could be used to lift each other from one level to the next to serve the work fronts, and then reversing the sequence to bring the cranes down after the structure and cladding were complete.

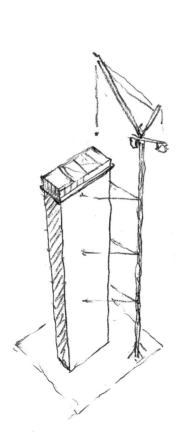

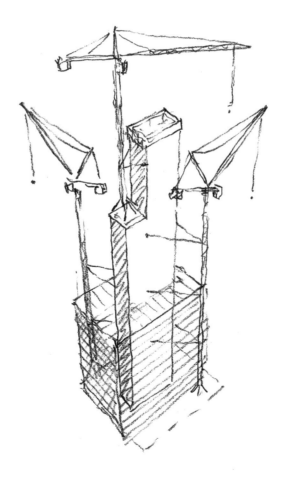

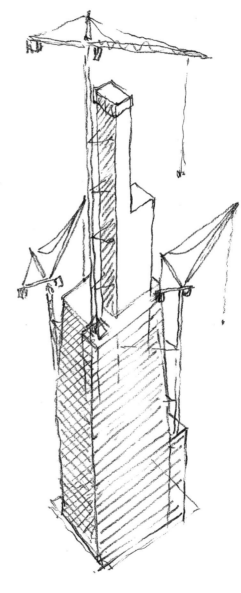

STAGE 1

STAGE 2

STAGE 3

SKETCHBOOK

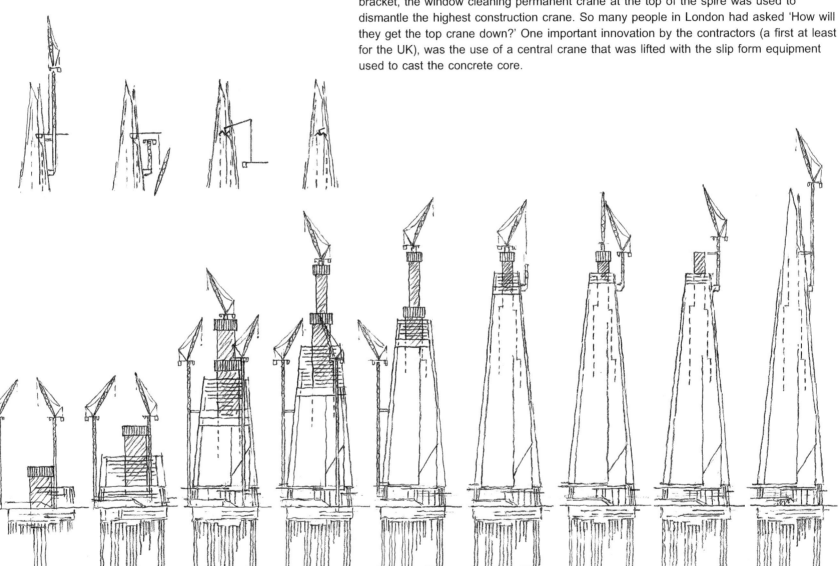

FIGURE 3.2/9

The Shard at London Bridge. These sequence sketches are a record of the actual craneage strategy used to construct the Shard. Several ideas developed during the initial design stages were adopted by contractors: cranes were lifted by each other to higher levels (the taper of the building meant that simple vertical extending masts could not be tied back safely to permanent structure at upper levels), a crane was built off a cantilever bracket, the window cleaning permanent crane at the top of the spire was used to dismantle the highest construction crane. So many people in London had asked 'How will they get the top crane down?' One important innovation by the contractors (a first at least for the UK), was the use of a central crane that was lifted with the slip form equipment used to cast the concrete core.

SKETCHBOOK

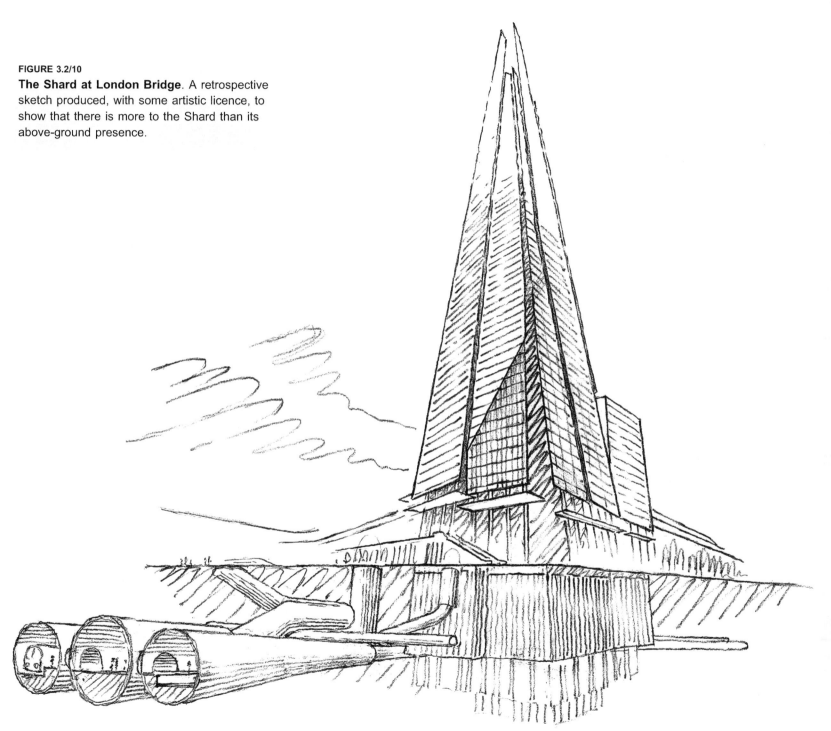

FIGURE 3.2/10

The Shard at London Bridge. A retrospective sketch produced, with some artistic licence, to show that there is more to the Shard than its above-ground presence.

FIGURE 3.2/11

Pinwheel, Beirut, Lebanon. A quick 'HB' pencil drawing works better in some circumstances to convey a single, simple message – here that cladding follows closely behind the structure. Consistent 'wobbly-line' quality is important and is best achieved by over-drawing very faint accurately set out guide lines – a not very sophisticated technique known as 'guided freehand'. This sketch was part of a construction sequence for RPBW's Pinwheel/ SOLIDERE scheme in Beirut.

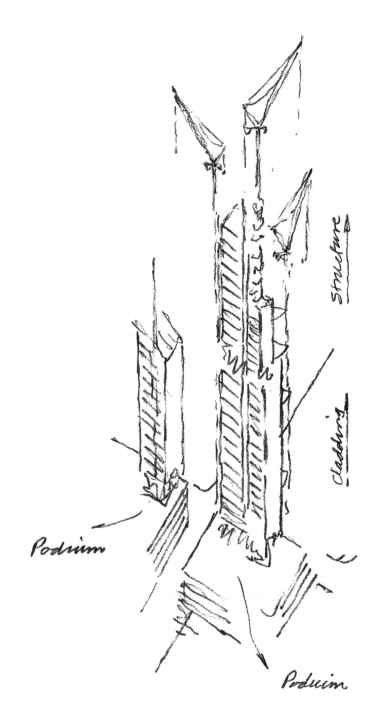

STAGE 4   Main superstructure

FIGURE 3.2/12

Al Burj. The foundation to the one-kilometre-high Al Burj was constructed when the recession started and put an end to work on site. It would have been the tallest building in the world and consisted of four discrete elements linked by sky-lobbies at three intermediate levels and at the top. The sub-division of the tower was a major factor with regard to aerodynamic performance and was also the basis of an important safety feature – transfer from one part of the building affected by an event to another unaffected part would be possible through the sky lobbies.

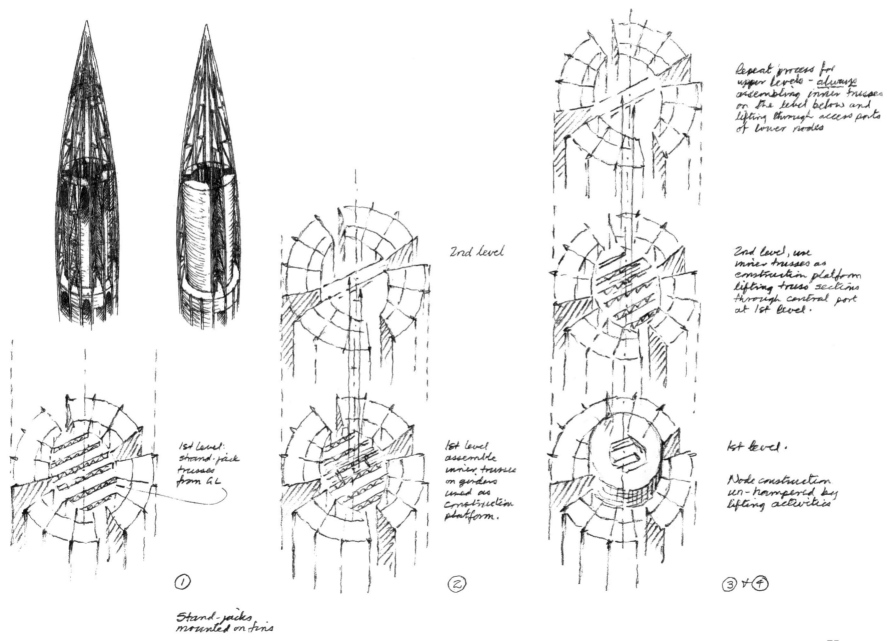

Repeat process for upper levels – always assembling inner trusses on the level below and lifting through access ports of lower nodes

2nd level

2nd level, use inner trusses as construction platform lifting truss sections through central port at 1st level.

1st level: strand-jack trusses from GL

1st level assemble inner trusses on girders used as construction platform.

1st level.

Node construction un-hampered by lifting activities

① ② ③ & ④

Stand-jacks mounted on fins

FIGURE 3.2/13

Project in Dubai with KPF. Proposals included cranes lifted up into position by each other, a crane located on a cantilevered bracket and a crane mounted on a slipform, all strategies used later on the Shard.

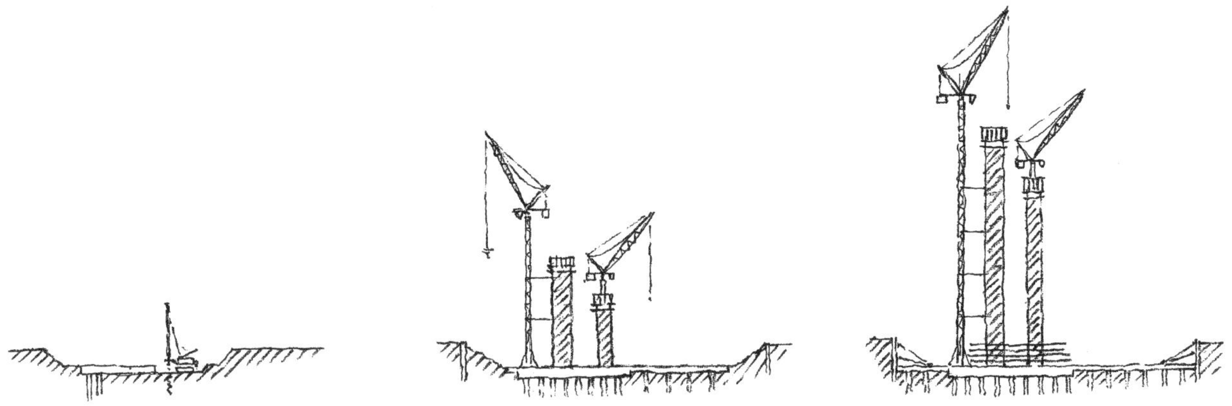

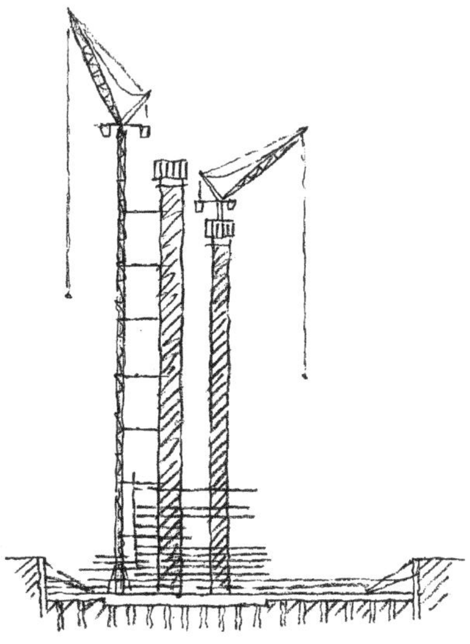

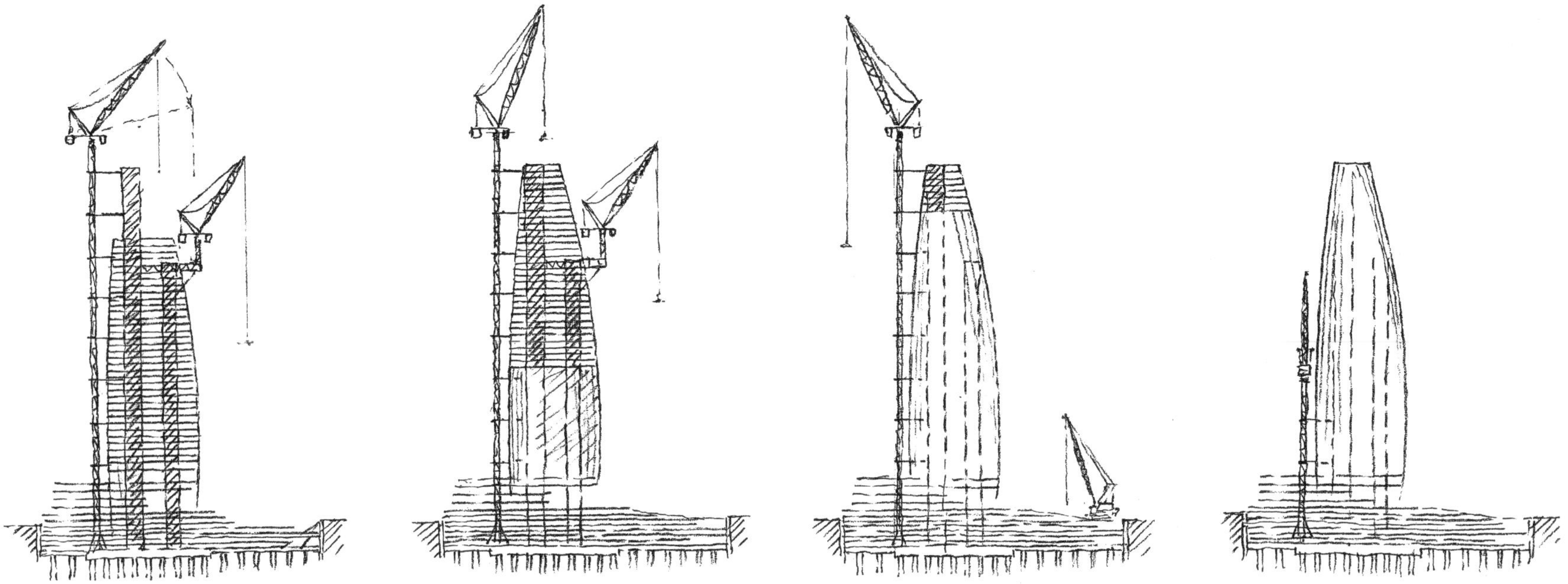

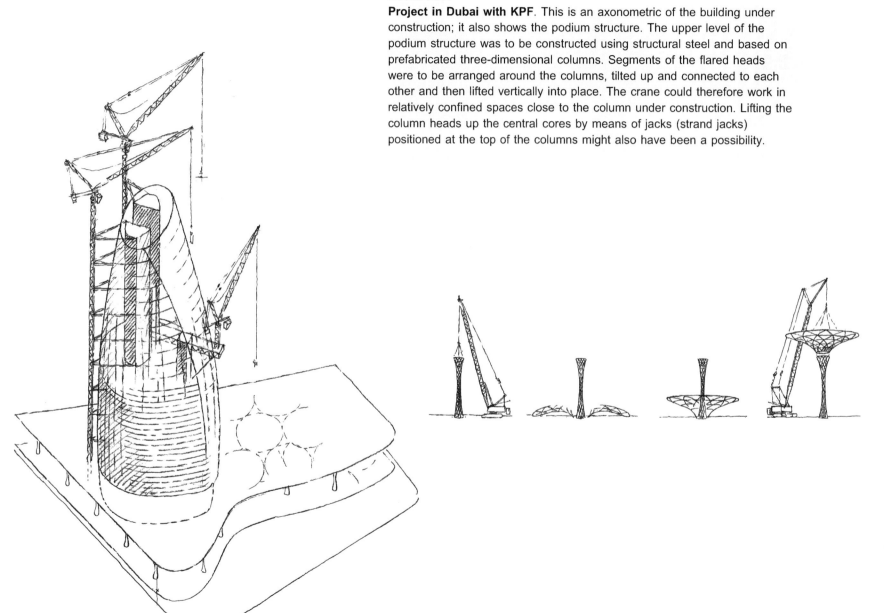

FIGURE 3.2/14

Project in Dubai with KPF. This is an axonometric of the building under construction; it also shows the podium structure. The upper level of the podium structure was to be constructed using structural steel and based on prefabricated three-dimensional columns. Segments of the flared heads were to be arranged around the columns, tilted up and connected to each other and then lifted vertically into place. The crane could therefore work in relatively confined spaces close to the column under construction. Lifting the column heads up the central cores by means of jacks (strand jacks) positioned at the top of the columns might also have been a possibility.

Bank headquarters, Middle East. Sometimes simple sketches are the most effective – limited use of colour would help and basic knowledge of tower crane type, scale and positioning provides a degree of credibility. Not all schemes are as easy as this one to describe visually, but for buildability and build sequencing it is always worth experimenting with two dimensional diagrams before investing time in anything more complicated.

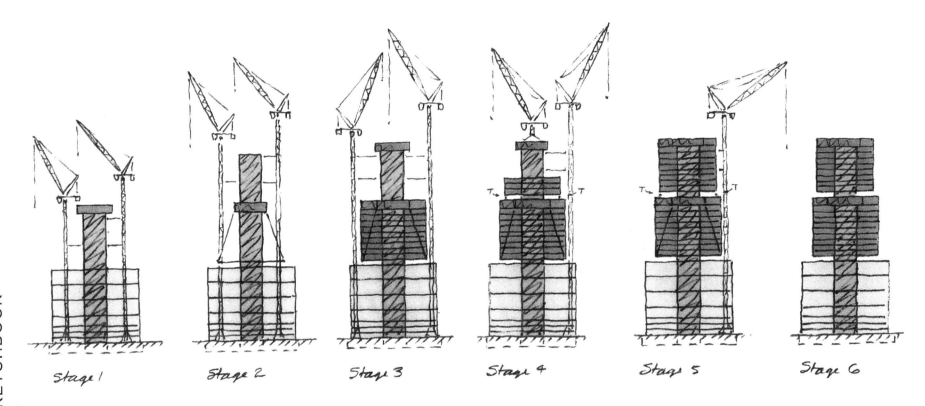

Stage 1   Stage 2   Stage 3   Stage 4   Stage 5   Stage 6

SKETCHBOOK

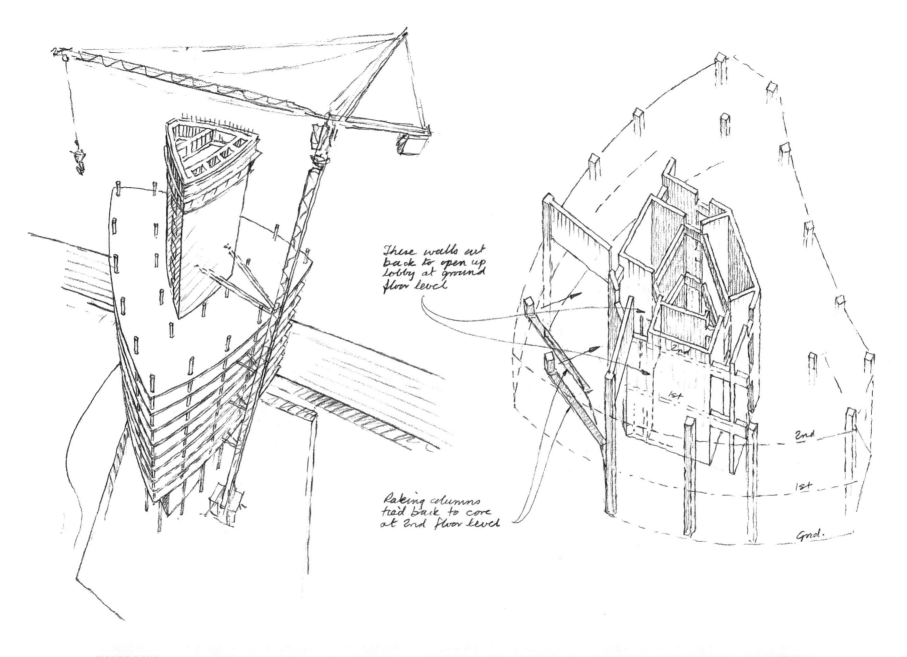

These walls cut back to open up lobby at ground floor level

Raking columns tied back to core at 2nd floor level

2nd

1st

2nd

1st

Gnd.

FIGURE 3.2/16

Strata, London. Strata is a very distinctive residential tower on the London skyline. The core was originally conceived as triangular on plan but eventually evolved into a more conventional rectangular configuration but assisted in terms of the lateral stability system by outrigger walls as shown here. Simple one point vertical perspective and axonometric core drawings were used at the formative stages of the project.

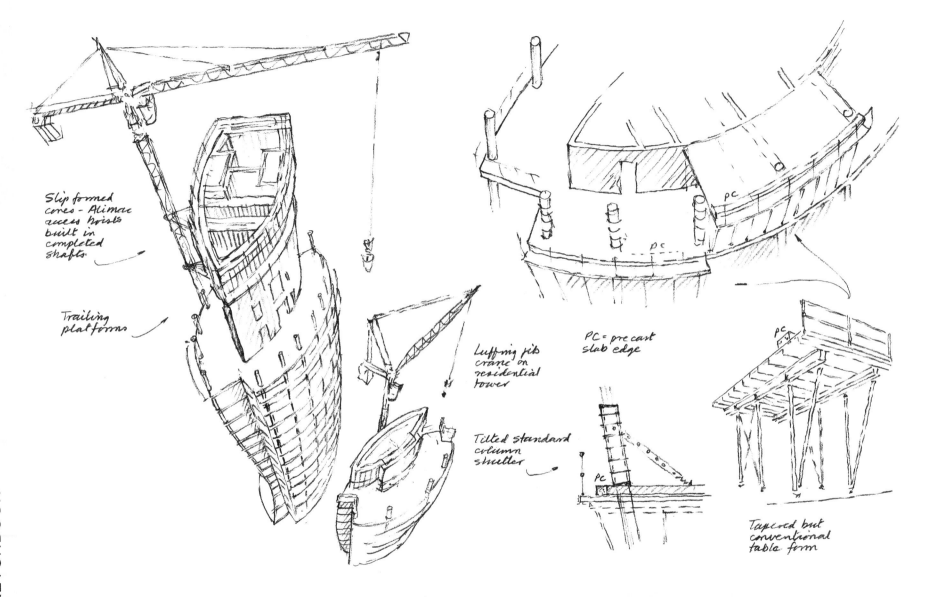

Slip formed cores - Alimac access hoists built in completed shafts.

Trailing platforms

Luffing jib crane on residential tower

Tilted standard column shutter

PC = pre cast slab edge

pc

Tapered but conventional table form

FIGURE 3.2/17
Twin towers, Kuwait. Vertical single point perspectives are surprisingly easy to do and good enough to summarise the basics of tower design. Details provide information on unusual aspects such as pre-cast slab edges and adjustable tilted shuttering for columns and tapered table formwork.

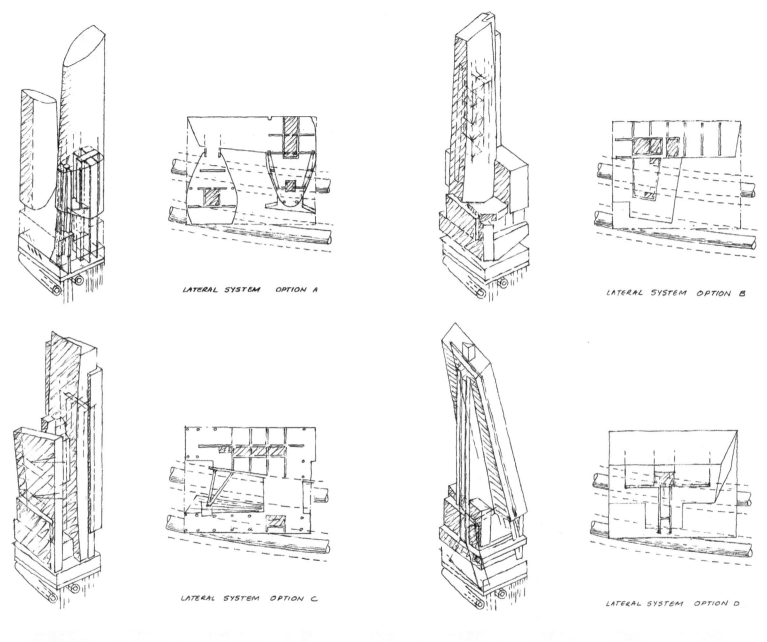

LATERAL SYSTEM OPTION A

LATERAL SYSTEM OPTION B

LATERAL SYSTEM OPTION C

LATERAL SYSTEM OPTION D

FIGURE 3.2/18

Columbus Tower, West India Quay. Columbus Tower was the name given to a scheme to be built over the Crossrail running tunnels at the western end of West India Quay in London's Docklands. The architect explored various massing options designed to accommodate a number of combinations of commercial, residential and retail space. Each proposal had to deal with the foundation constraints imposed by the tunnel alignments, tunnel exclusion zones and the existing, protected dock walls. In addition all the proposals were developed to include a viable lateral system, fundamental to the design of tall buildings – most here are based on shear walls on the major axis and out-riggered core walls on the minor axis.

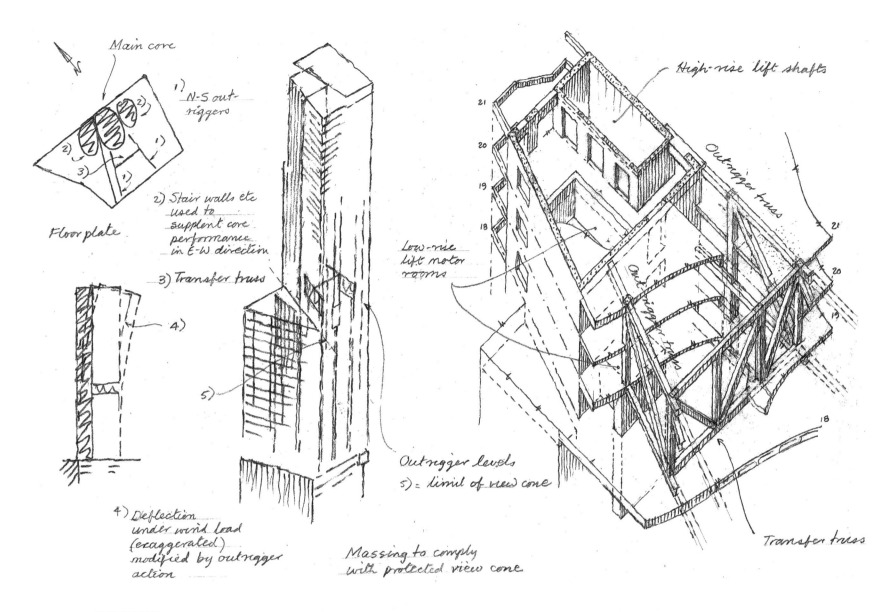

Main core

1) N-S outriggers

2) Stair walls etc used to supplent core performance in E-W direction

3) Transfer truss

Floor plate

4)

5)

4) Deflection under wind load (exaggerated) modified by outrigger action

High-rise lift shafts

Low-rise lift motor rooms

Outrigger levels
5) = limit of view cone

Massing to comply with protected view cone

Outrigger truss

Transfer truss

21
20
19
18

21
20
19
18

FIGURE 3.2/19

13–14 Appold Street, London. Appold Street is a world-class mixed use development located near Liverpool Street Station in the London Borough of Hackney. The building comprises forty-five storeys over a three-level basement on a non-rectilinear, relatively small footprint. A protected view corridor crosses the site and to a certain extent dictates the massing of the building. Mid-height steel trusses are designed to support columns in the upper level hotel floors so that the lower commercial floors can be column-free. The trusses also work as outriggers in conjunction with the core to provide a robust lateral stability system. The building is of necessity highly complex and versions of the sketch on the right, showing the transfer/outrigger system, were marked up to illustrate for example the use of different materials, how the stairs negotiate these complex levels and how the trusses are anchored back to the core. Sketches reproduced by kind permission of Masterworks Developments.

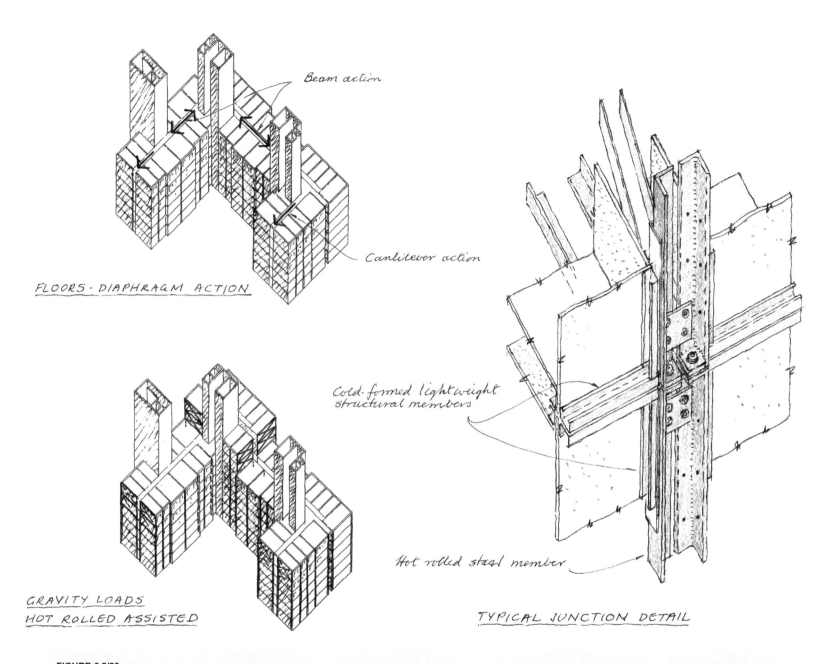

Beam action

Cantilever action

FLOORS - DIAPHRAGM ACTION

GRAVITY LOADS
HOT ROLLED ASSISTED

Cold-formed lightweight
structural members

Hot rolled steel member

TYPICAL JUNCTION DETAIL

FIGURE 3.2/20

High-rise modular. Cold formed light gauge modular construction consisting of factory produced modules had been used for some time in low rise projects, especially for hotels and student accommodation. These sketches were part of a study to develop a high-rise system based on modular units but with alternative lateral stability systems. Upper left relies on slip formed reinforced concrete cores designed as the primary stability elements and with the floor plates providing horizontal diaphragm action. Bottom left shows a different system where the lateral loads are carried by hot-rolled steel shear frames and the cores are constructed from lightweight panels. Detailing, especially at junctions, was the key to success of this project.

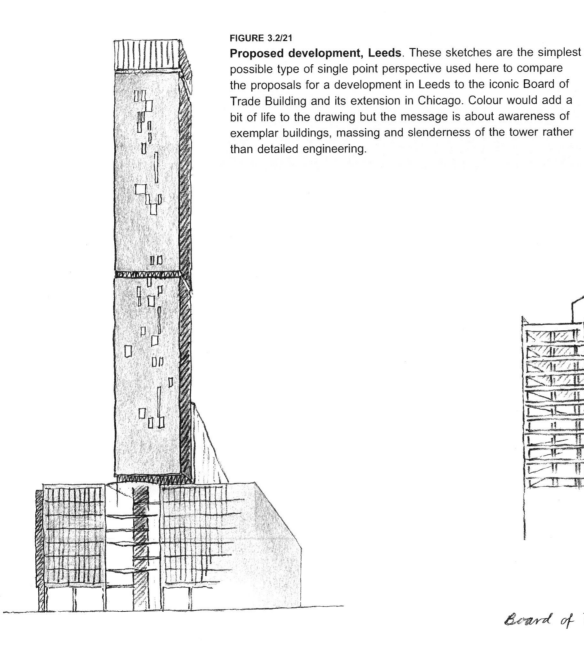

FIGURE 3.2/21

Proposed development, Leeds. These sketches are the simplest possible type of single point perspective used here to compare the proposals for a development in Leeds to the iconic Board of Trade Building and its extension in Chicago. Colour would add a bit of life to the drawing but the message is about awareness of exemplar buildings, massing and slenderness of the tower rather than detailed engineering.

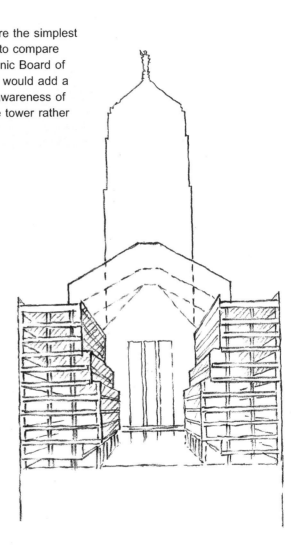

Board of Trade Building Chicago Extension

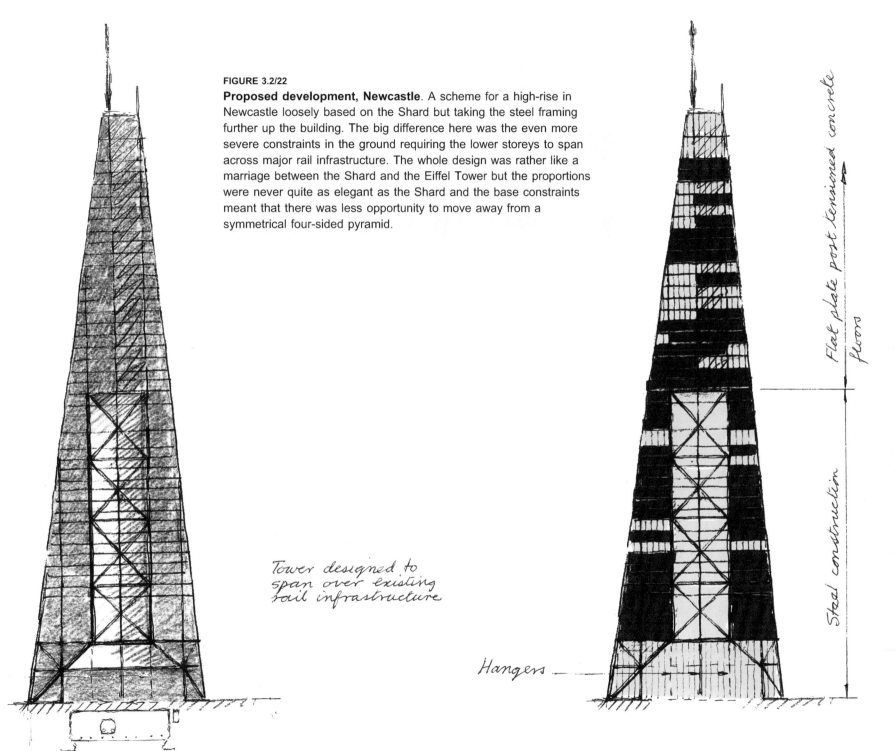

FIGURE 3.2/22

Proposed development, Newcastle. A scheme for a high-rise in Newcastle loosely based on the Shard but taking the steel framing further up the building. The big difference here was the even more severe constraints in the ground requiring the lower storeys to span across major rail infrastructure. The whole design was rather like a marriage between the Shard and the Eiffel Tower but the proportions were never quite as elegant as the Shard and the base constraints meant that there was less opportunity to move away from a symmetrical four-sided pyramid.

Tower designed to span over existing rail infrastructure

Hangers

Flat plate post tensioned concrete floors

Steel construction

FIGURE 3.2/23

Twisting Tower, Dublin. Another victim of the 2007/8 recession, the Twisting Tower on the Dublin waterfront would have been an imposing building. Here the main design issue was how to fit the various duplex apartment layouts together. This was first solved as if the stack was vertical, and then creating the spiral by stepping the walls gradually. Gravity loads were carried on sloping perimeter columns

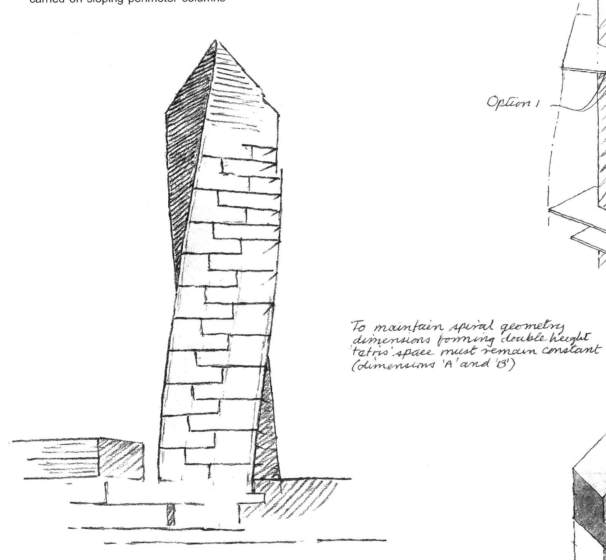

Option 2

Option 1

Best solution..

Space optimised on every level

Perimeter columns on on these lines

To maintain spiral geometry dimensions forming double height 'tetris' space must remain constant (dimensions 'A' and 'B')

This leads to two options: Option 1 'step' walls radiate from corners Option 2 'step' walls remain at 90° to facades (cut out shape is un-altered)

FIGURE 3.2/24

Hotel, Doha. Designer and engineer have to work closely together to produce practical and buildable solutions for initial concepts that are in some ways sculptural – the functionality of the building and then the structure have to be fitted into the original design, as per this futuristic scheme by Heatherwick Studio for the Grand Hotel in Doha.

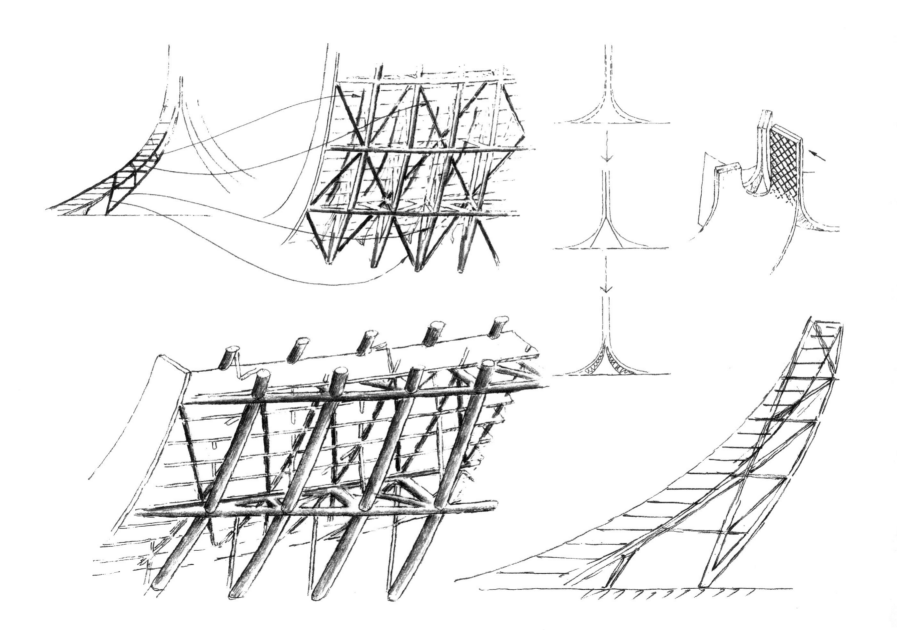

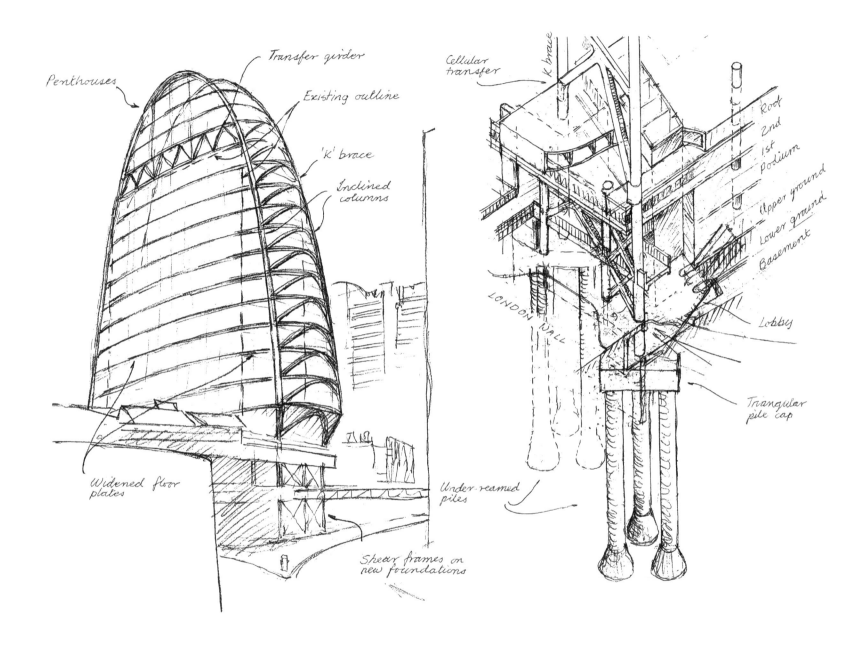

Penthouses

Transfer girder

Existing outline

'K' brace

Inclined columns

Cellular transfer

K brace

Roof
2nd
1st
Podium
Upper ground
Lower ground
Basement

LONDON WALL

Lobby

Triangular pile cap

Widened floor plates

Under-reamed piles

Shear frames on new foundations

FIGURE 3.2/25
Bastion House, London Wall, London. Plans to redevelop Bastion House in the City of London based on a partial rebuild never progressed beyond concept stage. Sketches such as these were used to explain the complexities of the scheme, especially with regard to the feasibility of constructing new foundations required to support a new frame built over and around the existing structure.

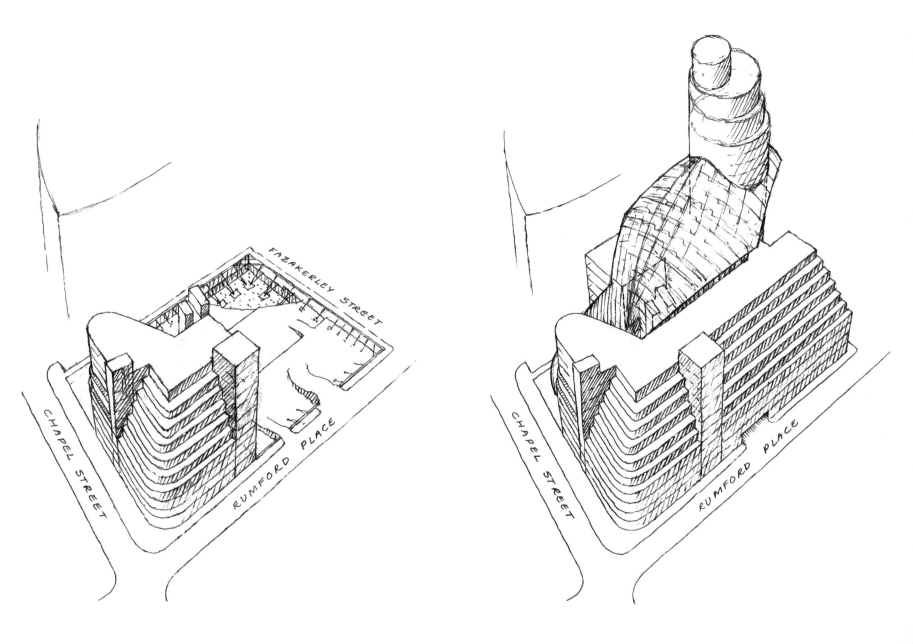

FIGURE 3.2/26

Proposed development, Liverpool. This was a design competition to redevelop an important site not too far from the waterfront and some of Liverpool's iconic buildings. Simple axonometric sketches show the stepped façades of the first construction phase and the massing for the second phase atrium and tower.

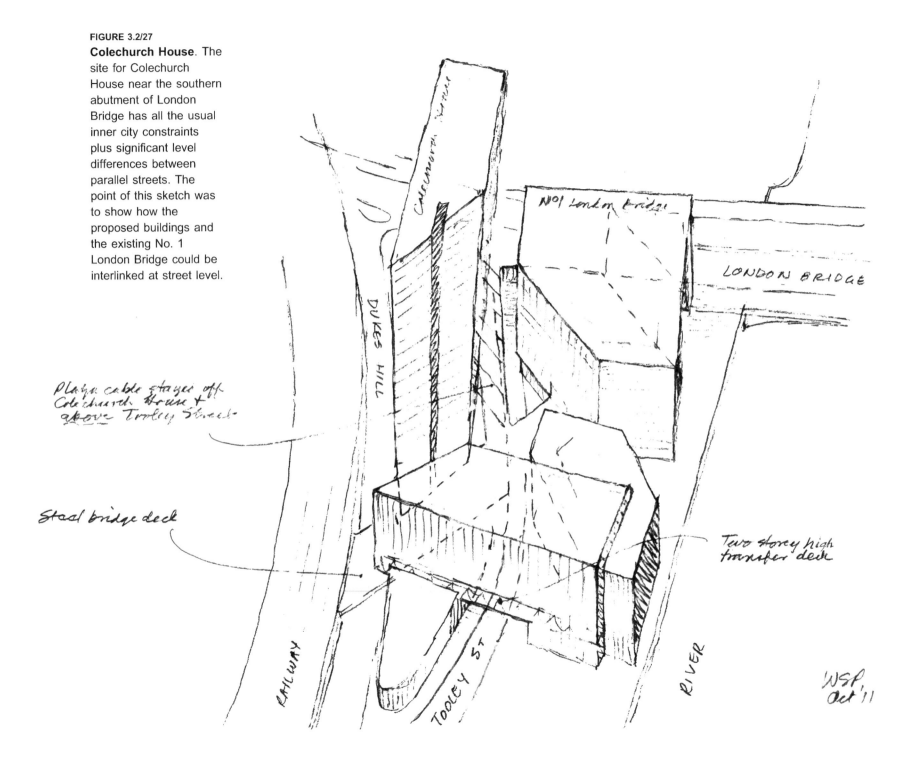

FIGURE 3.2/27

Colechurch House. The site for Colechurch House near the southern abutment of London Bridge has all the usual inner city constraints plus significant level differences between parallel streets. The point of this sketch was to show how the proposed buildings and the existing No. 1 London Bridge could be interlinked at street level.

Colechurch House

N°1 London bridge

LONDON BRIDGE

DUKES HILL

Plaza cable stayed off Colechurch House + above Tooley Street

Steel bridge deck

Two storey high transfer deck

RAILWAY

TOOLEY ST

RIVER

WSP Oct '11

FIGURE 3.2/28

Milton Court, Moor Lane, City of London. Milton Court, a major development in the City of London, contains a world class concert hall, a teaching theatre and a residential tower. Spatial requirements for the various uses were fitted together rather like a three dimensional puzzle on this highly constrained site. This cutaway section shows the teaching theatre which is positioned directly below the thirty-six storey residential tower and shows the relationship between the tower core, shear walls and transfer structure carrying load across the theatre itself.

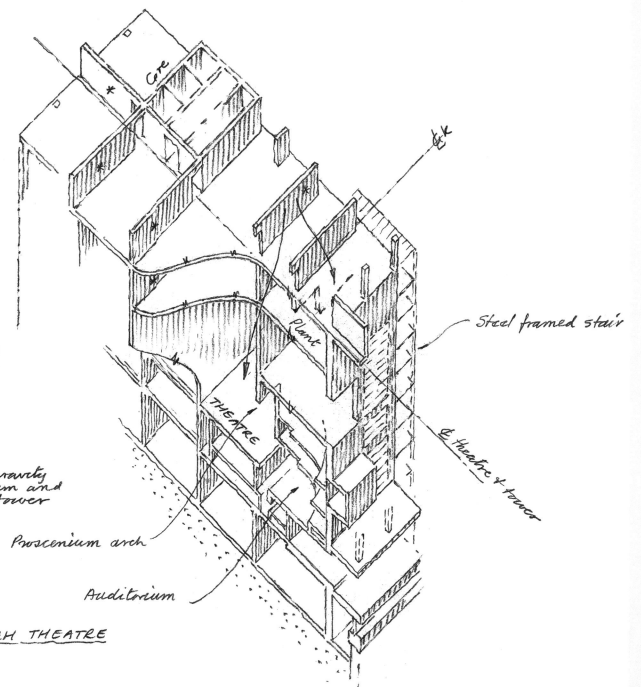

Core

Plant

Steel framed stair

THEATRE

℄ theatre & tower

℄ K

* Remove these walls from original design

Walls on ℄ K transfer gravity loads around auditorium and torsional restraint to tower

Proscenium arch

Auditorium

CUTAWAY SECTION THROUGH THEATRE

SKETCHBOOK

91

FIGURE 3.2/29

Milton Court, Moor Lane, City of London. Sequence sketches were produced for Milton Court to show different design solutions adopted to take account of various site perimeter conditions. A secant wall was used where gravity loads from the new development are high and a king post embedded wall was used on the western boundary where loads are much lower. The southwest was particularly difficult due to the existence of complex buried services.

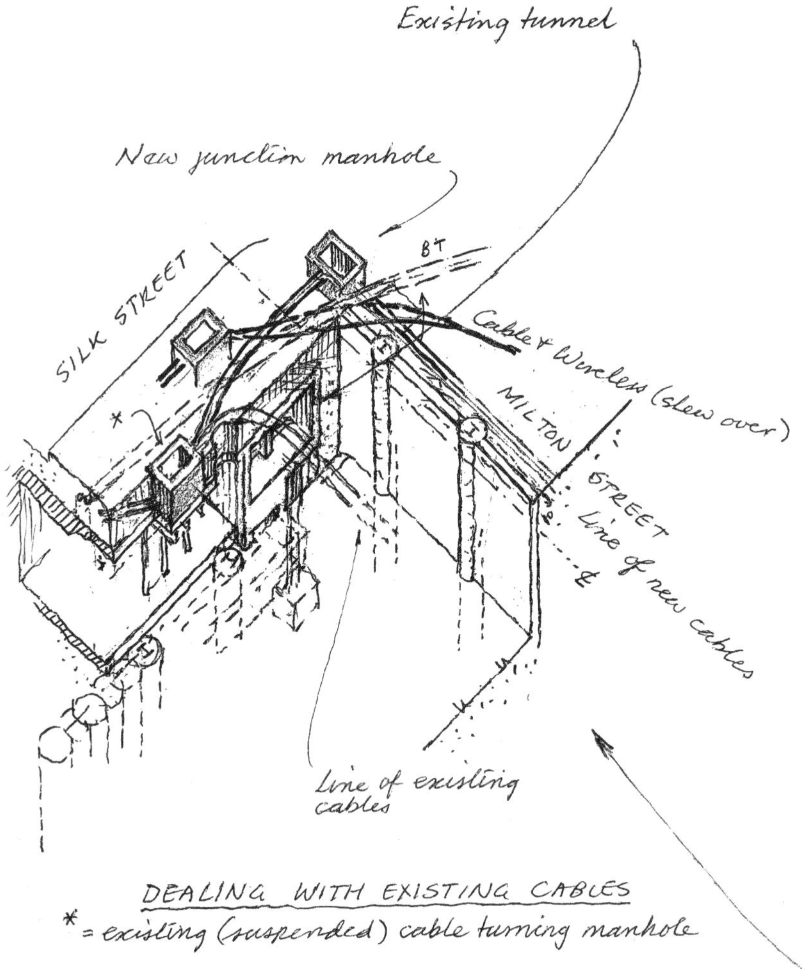

Existing tunnel

New junction manhole

SILK STREET

BT

Cable & Wireless (slew over)

MILTON STREET

Line of new cables

Line of existing cables

DEALING WITH EXISTING CABLES
* = existing (suspended) cable turning manhole

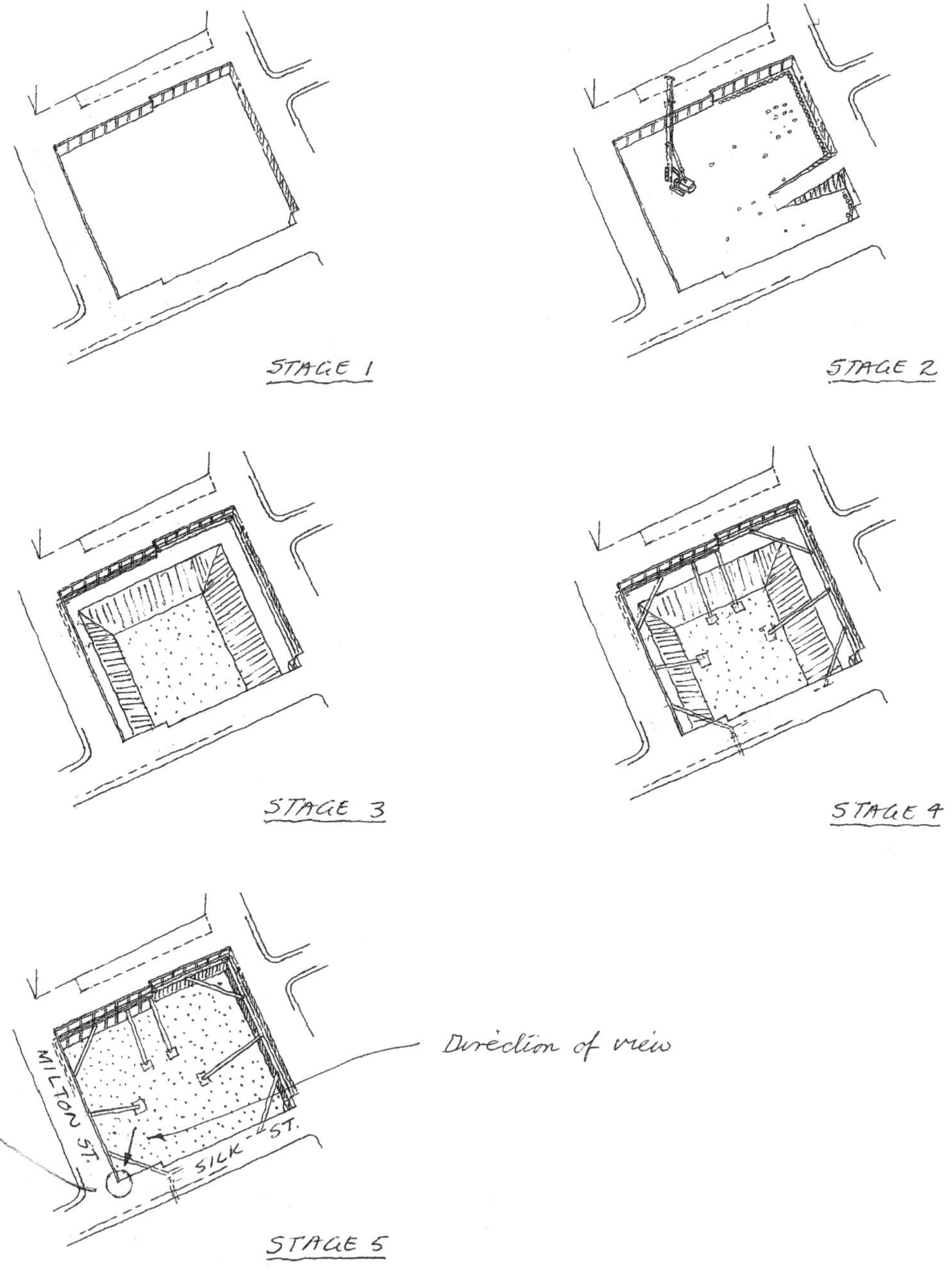

STAGE 1

STAGE 2

STAGE 3

STAGE 4

STAGE 5

MILTON ST.

SILK ST.

Direction of view

SKETCHBOOK

92

FIGURE 3.2/30

Leadenhall Triangle, Billiter Street, London. The City of London is particularly complex, with crowded buildings separated by narrow lanes and alleyways. Understanding the site context is important during the initial planning stages, and here at Leadenhall Triangle the development proposals are heavily influenced by boundary conditions and the existence of a listed building in a rather inconvenient position facing Billiter Street.

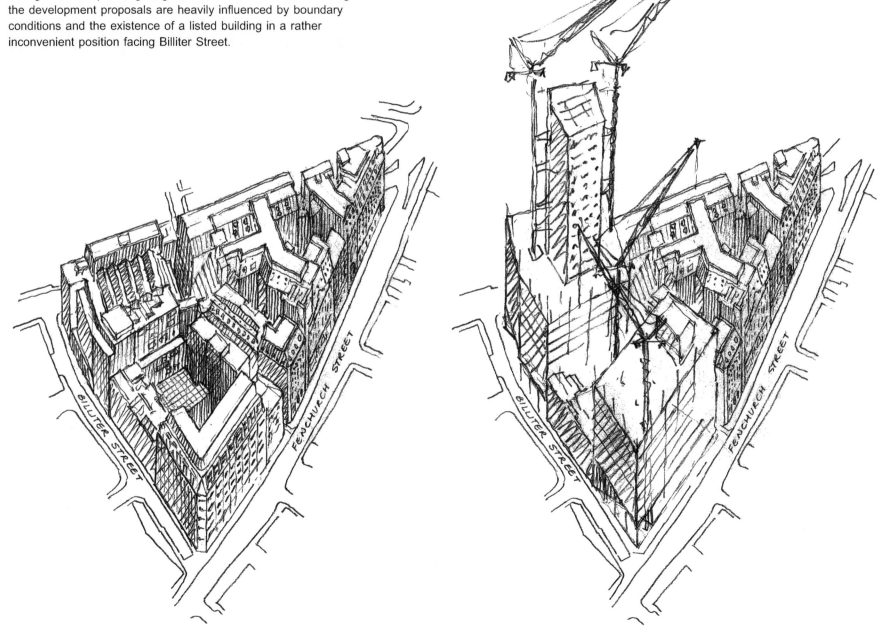

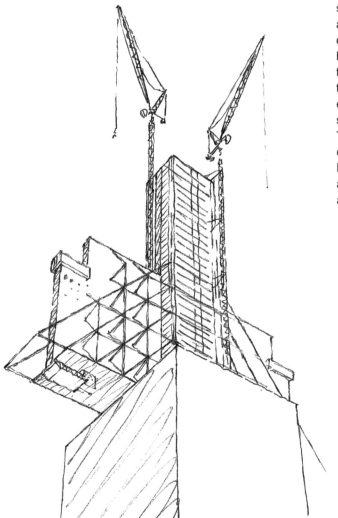

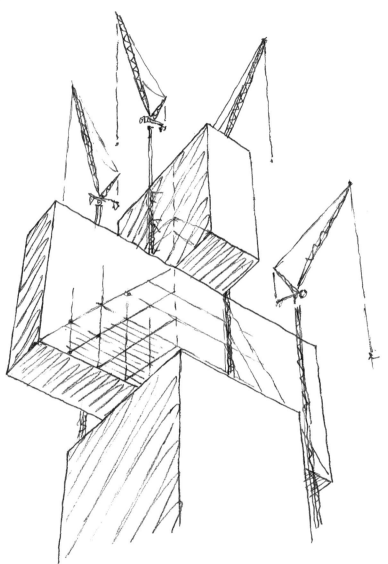

FIGURE 3.2/31
New Jersey, USA. This was a scheme that never developed beyond the first simplistic idea of stacking blocks one above the other. Each cantilever wing would have been built out from the central core using the full depth of the elevation to provide strength and stiffness. The core alone would carry lateral loads. Extreme engineering for a rather ambitious architectural concept.

3.3   LOW- AND MID-RISE

Low- and mid-rise is really a catch-all category and probably covers 90 per cent of buildings in most places, ranging from new build to refurbishment, from 5 or 6 storey offices, to residential blocks, art galleries and supermarkets. Nevertheless, they bring many of the same issues and challenges that apply to larger, taller or more specialised buildings. Sometimes they are less reliant on grand engineering solutions and more on careful, crisp detailing. Good clear, clean detailing isn't easy to achieve; simplicity and neat solutions take time to develop and require a great deal of interdisciplinary cooperation and effort. Many low- or mid-rise buildings have become landmarks or are considered exceptional pieces of design. Whether a project achieves this status or not, there is still a lot of satisfaction to be had in delivering a well-thought-out end product.

Take supermarkets for instance, at first sight an uninspiring building type. But the whole sector has its special challenges and is very competitive so there is a need to build economical and efficient structures. The pressure often focuses on designing a simple, lightweight structure while minimising the number of columns, particularly in the sales area. Sustainability objectives based on the use of natural ventilation and natural light, on the use of recycled materials and reduced embodied energy are becoming normal practice. Shorter and shorter construction programmes are also driving the buildability aspects of the design approach. Over and above all of this, architects find these projects equally challenging and will fight hard to make their mark and do something that hasn't been done before.

Sketching on a supermarket project is often needed to explain a proposed construction sequence to a vast array of interested parties ranging from clients to designers, programmers and contractors. Just now and again, someone will propose a flagship store or a radical design that will need special attention and provide great opportunities to do some fun sketches.

Sketching on refurbishment or alteration projects is different. Here it may be important to pick out and illustrate particular areas only and leave other parts in sketchy or broad outline. It is often as important to decide what not to show as it is to show the important features. Cutaway illustrations come into their own on refurbishment and alteration projects.

Hand sketching also lends itself to communication on domestic scale projects, because it is quick to do and easy for non-technical clients to understand. This applies to rear extensions and to palaces for Russian oligarchs alike.

Low- and mid-rise buildings may not have the instantaneous attraction of other categories, but every single construction project will have its own challenges and opportunities.

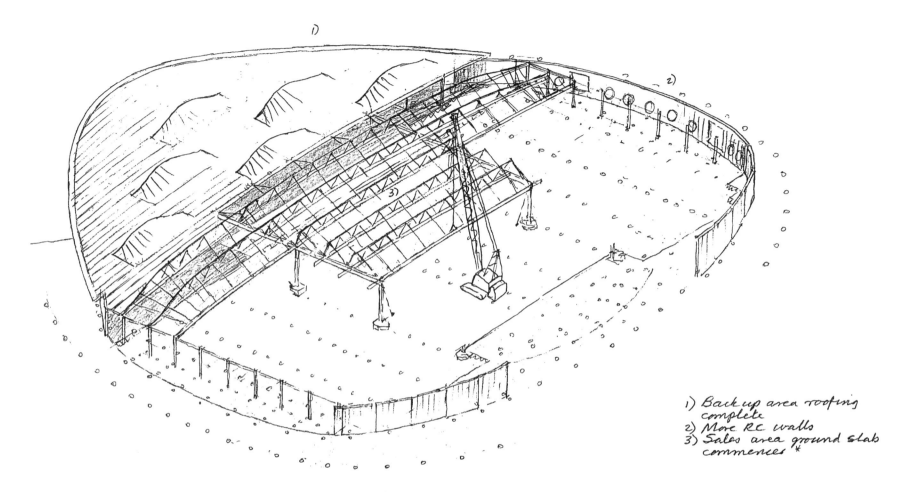

1) Back up area roofing complete
2) More RC walls
3) Sales area ground slab commences *

* Sales area ground slab.- 150 thk lowest slab constructed on gas membrane etc, designed for light construction loads only (as ground bearing slab in short term - as suspended slab for service loading only for long term). Dwarf walls cast later using pull-out starter bars - in meantime, flat slab allows free access for roofing activities.

FIGURE 3.3/1

J. Sainsbury, Greenwich, London. Sainsbury's millennium store at Greenwich won fourteen awards for sustainable design. The main features were: natural light, natural ventilation, use of recycled materials, ground source heating and cooling, and the use of combined heat and power plant. The structure was based on 'reasonable' spans allowing the design of efficient structure further reducing embodied energy. Sketch reproduced by kind permission of Sainsbury's Supermarkets Ltd.

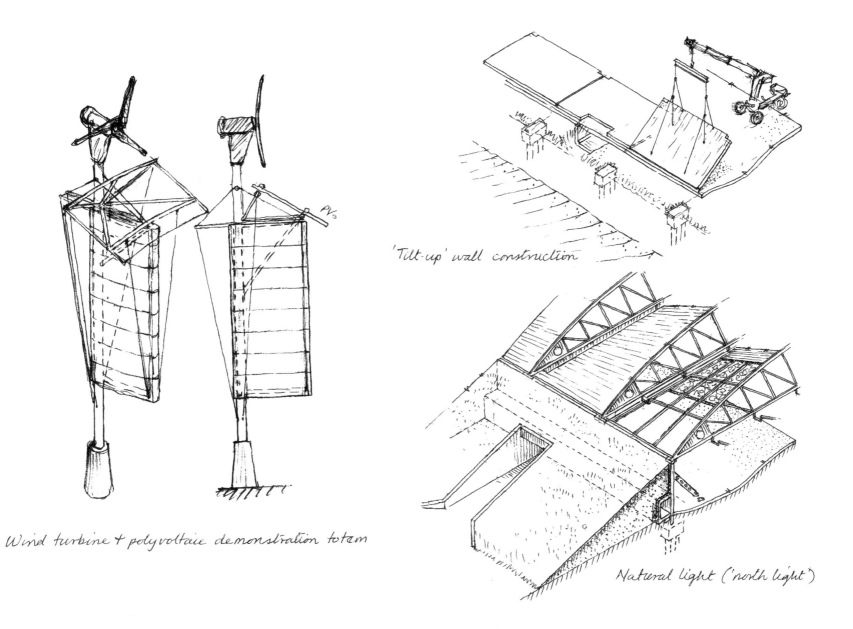

'Tilt-up' wall construction

Wind turbine + polyvoltaic demonstration totem

Natural light ('north light')

FIGURE 3.3/2

J. Sainsbury, Greenwich, London. The raised floor was designed to draw fresh air from outside through perimeter earth embankments into the sales area, where stack effect causes it to rise towards ventilation in the apexes of the north-light roof voids. Rather traditional north-light roof configuration allowed maximum penetration of good quality natural light. However, it has to be admitted that the 'tilt-up' method of casting perimeter wall panels on the ground (top right) and rotating them into place (a popular North American and Australian technique) never happened and the wind turbines were no more than tokenism. Sketch reproduced by kind permission of Sainsbury's Supermarkets Ltd.

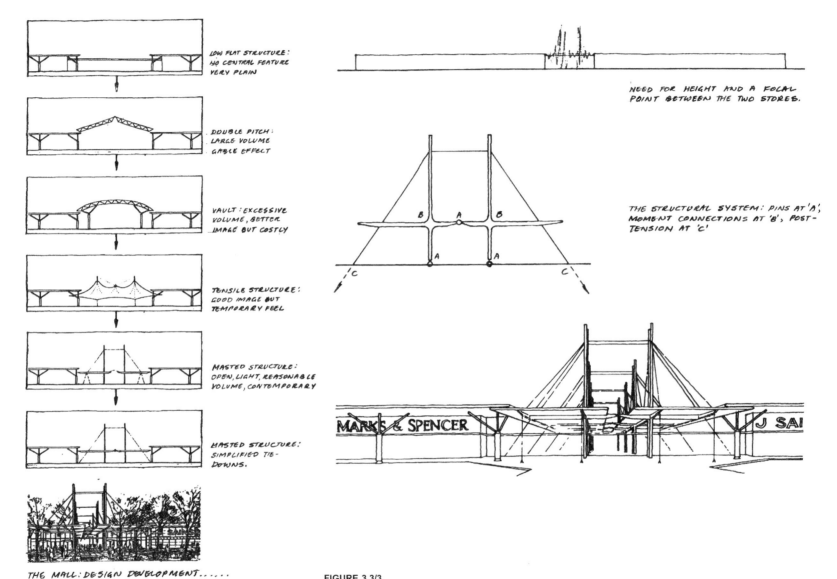

LOW FLAT STRUCTURE: NO CENTRAL FEATURE VERY PLAIN

DOUBLE PITCH: LARGE VOLUME GABLE EFFECT

VAULT: EXCESSIVE VOLUME, BETTER IMAGE BUT COSTLY

TENSILE STRUCTURE: GOOD IMAGE BUT TEMPORARY FEEL

MASTED STRUCTURE: OPEN, LIGHT, REASONABLE VOLUME, CONTEMPORARY

MASTED STRUCTURE: SIMPLIFIED TIE-DOWNS.

THE MALL: DESIGN DEVELOPMENT......

NEED FOR HEIGHT AND A FOCAL POINT BETWEEN THE TWO STORES.

THE STRUCTURAL SYSTEM: PINS AT 'A', MOMENT CONNECTIONS AT 'B', POST-TENSION AT 'C'

MARKS & SPENCER   J SAI

FIGURE 3.3/3

Marks & Spencer and Sainsbury's at Hedge End, Hampshire. Something special had to be designed to act as a focal point and announce the entrances to the Marks & Spencer and Sainsbury's stores at Hedge End. Many forms were examined but the final choice rested on the perceived need for height rather than on any structural logic. Sketch reproduced by kind permission of Sainsbury's Supermarkets Ltd.

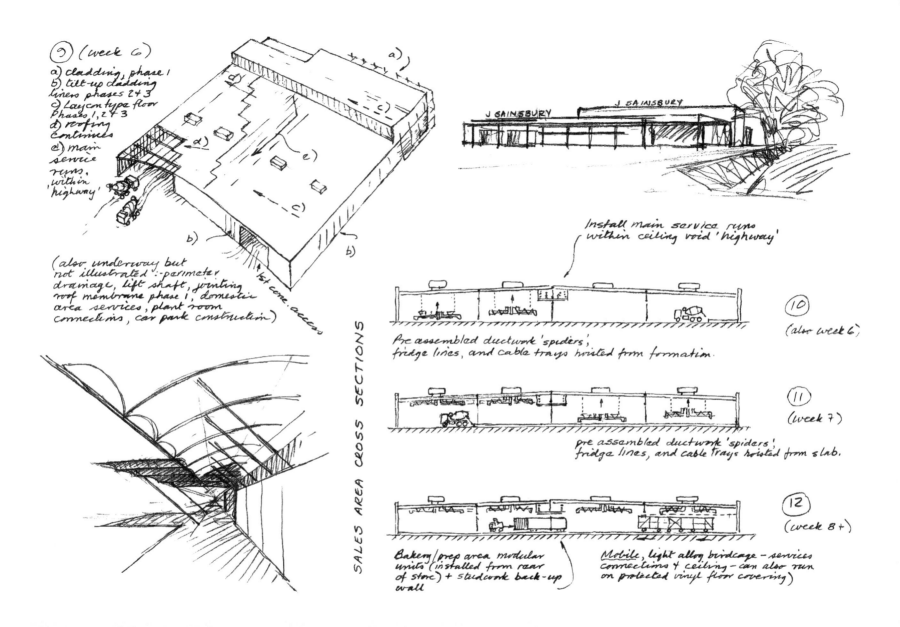

9 (week 6)
a) cladding, phase 1
b) tilt-up cladding liners phases 2 + 3
c) Laycon type floor Phases 1, 2 + 3
d) roofing continues
e) main service runs within 'highway'

(also underway but not illustrated:- perimeter drainage, lift shaft, jointing roof membrane phase 1, domestic area services, plant room connections, car park construction)

1st conc. access

J SAINSBURY   J SAINSBURY

Install main service runs within ceiling void 'highway'

SALES AREA CROSS SECTIONS

10 (also week 6)
Pre assembled ductwork 'spiders', fridge lines, and cable trays hoisted from formation.

11 (week 7)
pre assembled ductwork 'spiders', fridge lines, and cable trays hoisted from slab.

12 (week 8+)
Bakery/prep area modular units (installed from rear of store) + studwork back-up wall

Mobile, light alloy birdcage – services connections + ceiling – can also run on protected vinyl floor covering)

FIGURE 3.3/4

J. Sainsbury, the Dome Roundabout, Watford. Another Sainsbury supermarket – these sketches showed ideas to streamline construction and shorten build duration but the real challenge was to maintain the simple, clean elegant lines of the building. Crisp, clear detailing to match the overall simplicity isn't achieved easily and takes an inordinate amount of effort from the whole design team. Sketch reproduced by kind permission of Sainsbury's Supermarkets Ltd.

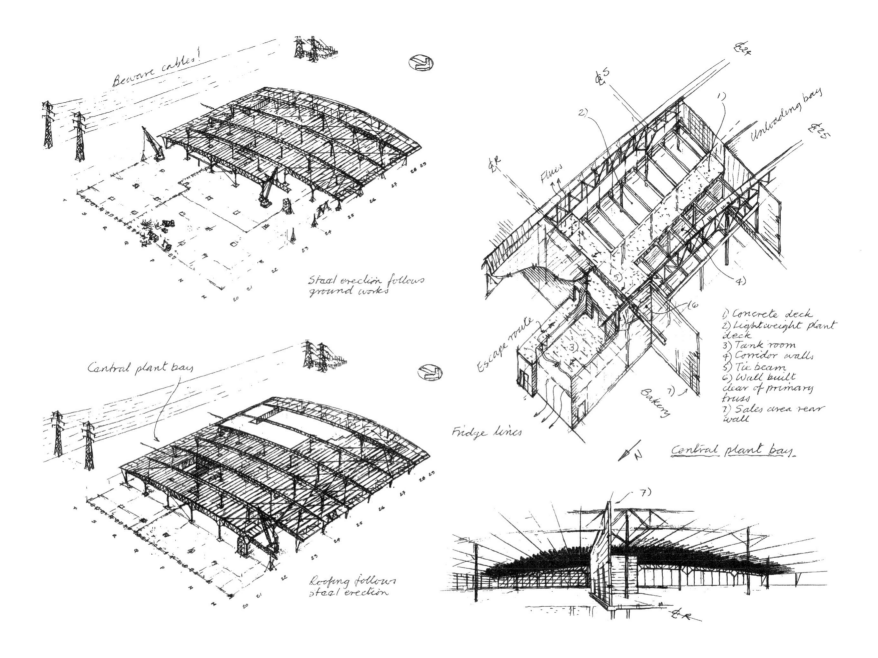

Beware cables!

Steel erection follows
ground works

Central plant bay

Roofing follows
steel erection

Flues

Escape route

Fridge lines

Unloading bay

Bakery

1) Concrete deck
2) Lightweight plant
deck
3) Tank room
4) Corridor walls
5) Tie beam
6) Wall built
clear of primary
truss
7) Sales area rear
wall

N

Central plant bay.

FIGURE 3.3/5

Savacentre Superstore, Beckton Triangle, London. Savacentre for Sainsbury's at Beckton was designed to be economic. The scheme was based on reasonably spaced columns on a regular grid supporting continuous lightweight steel trusses. Ground conditions were poor but suitable for driven piles which were used to support the frame and the ground slab. From an engineering point of view, the unclad structure looked fantastic: it was a real shame to hide it behind ceilings and roofing. Sketch reproduced by kind permission of Sainsbury's Supermarkets Ltd.

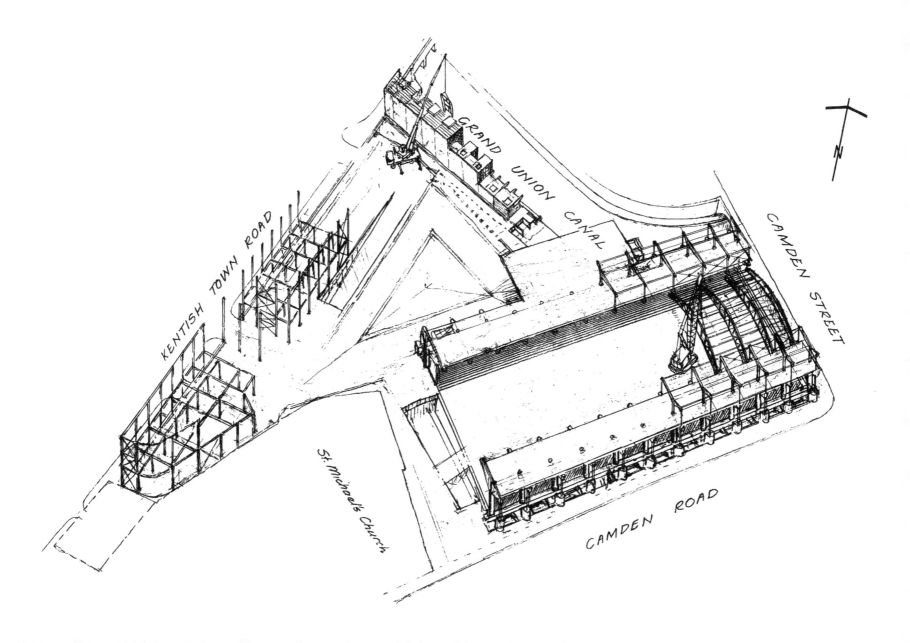

FIGURE 3.3/6

J. Sainsbury, Camden Town, London. In the mid-1980s this was a controversial design and even now divides opinion. The main building was set to match the surrounding streetscape in height and grain but there any reference to historical context ended. It was without question a very different 'high-tech' approach to supermarket design: exposed and expressed structure and a clear span providing column free space for future flexibility. It followed the London tradition of placing new architecture alongside older styles. Sketch reproduced by kind permission of Sainsbury's Supermarkets Ltd.

SKETCHBOOK

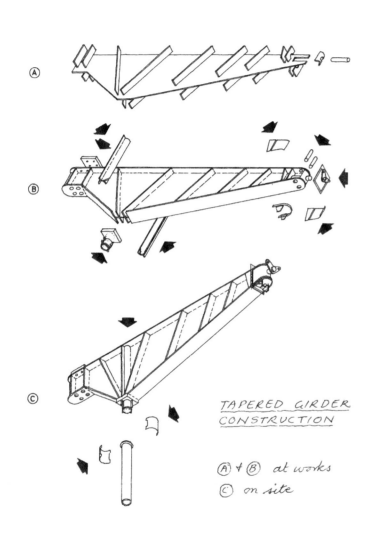

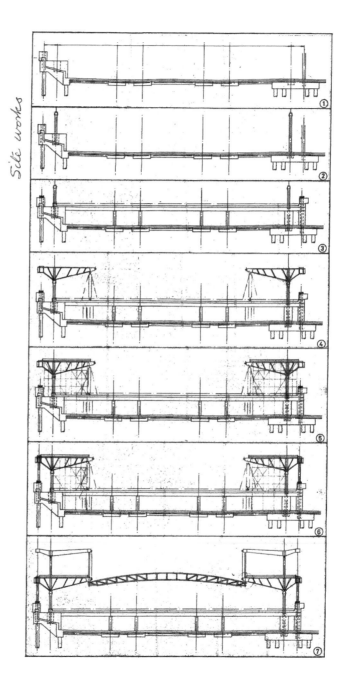

(A)

(B)

(C)

TAPERED GIRDER
CONSTRUCTION

(A) + (B) at works
(C) on site

Site works

①
②
③
④
⑤
⑥
⑦

FIGURE 3.3/7
J. Sainsbury, Camden Town, London. The design of Nicholas Grimshaw's unique supermarket at Camden Town is based on two-storey back-stayed perimeter structure supporting an internal simply supported roof truss. These hand drawings just preceded widespread use of CAD and are 'guided freehand' sketches (key lines drawn faintly using a drawing board and set squares). They show how the tapered girders were put together and the erection sequence. Sketch reproduced by kind permission of Sainsbury's Supermarkets Ltd.

FIGURE 3.3/8

The Barry Rooms, the National Gallery, London. Old roofs over a central group of galleries known as the Barry Rooms were replaced as part of a major refurbishment project at the National Gallery. New clear spanning structure was designed to allow free travel of cleaning gantries used to provide safe access to the lay-light glass immediately above the galleries. The cutaway drawing shows the rotating gantry suspended from structure spanning over the central dome.

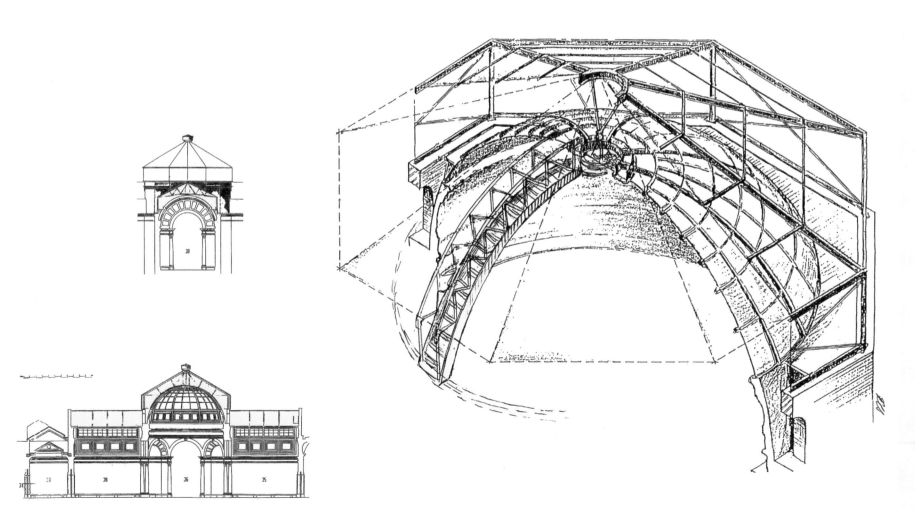

The National Portrait Gallery, London. This is a construction sequence sketch for a neat little scheme to infill a courtyard between the National Portrait Gallery and the National Gallery. The infill was an extension to the National Portrait Gallery and consequently the boundary only of the National Gallery is shown. Colour was used to separate proposed construction from existing structure.

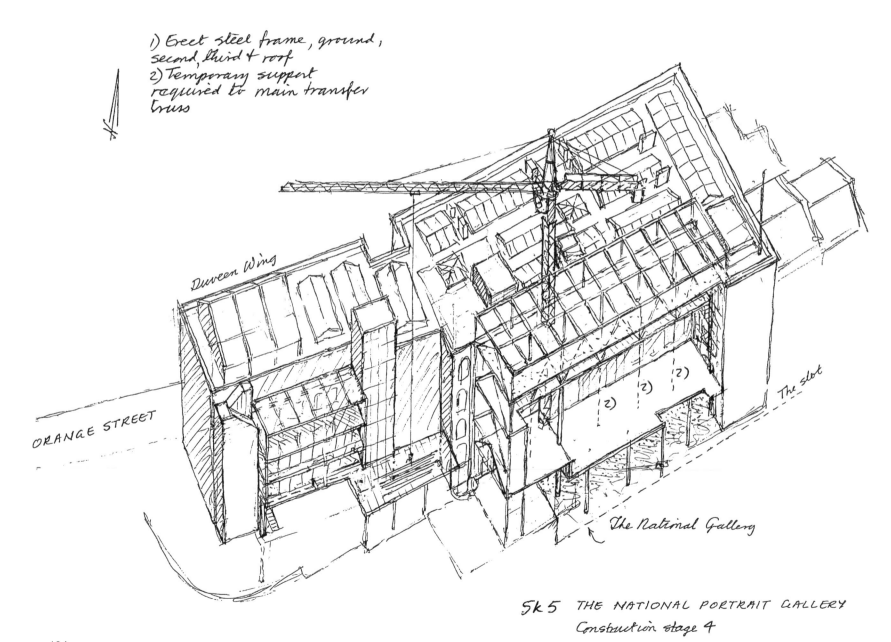

1) Erect steel frame, ground, second, third + roof
2) Temporary support required to main transfer truss

Duveen Wing

ORANGE STREET

The slot

The National Gallery

Sk 5 THE NATIONAL PORTRAIT GALLERY
Construction stage 4

SKETCHBOOK

Arts building, Manchester. The form of this proposal by RHWL Arts Team for the Manchester First Street arts building (subsequently called 'Home') is based on interconnecting boxes. Complexity is concentrated at the front of the building in the foyer area, and here the steel framing is arranged to be as unobtrusive as possible. Externally the massing is clean and simple and doesn't reveal anything about how the building is structured.

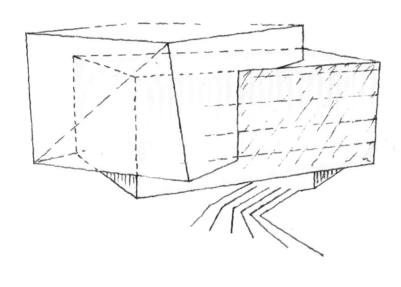

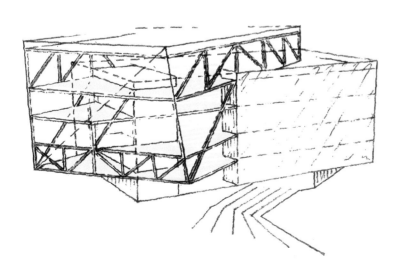

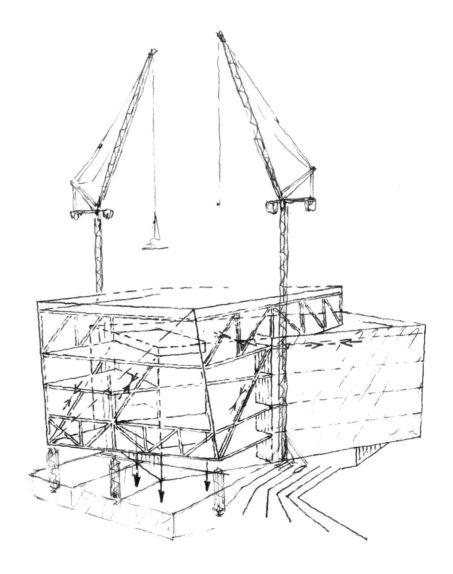

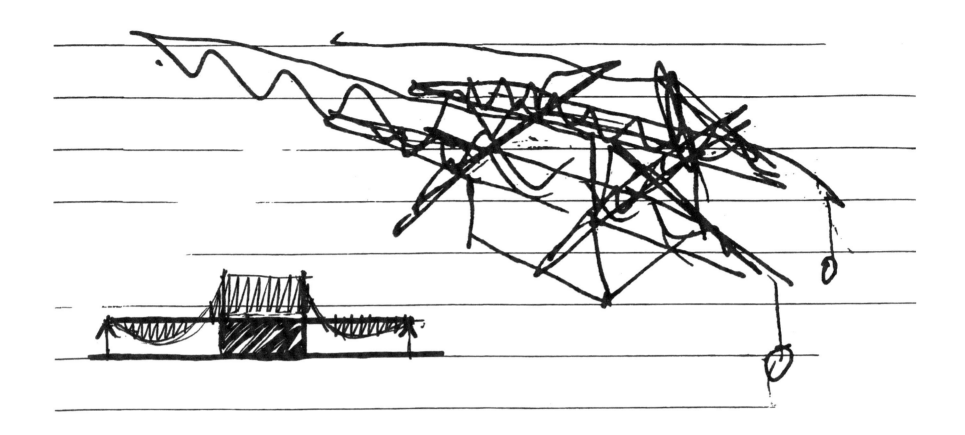

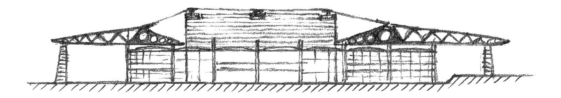

FIGURE 3.3/11

BRIT School for the Performing Arts, London. From the outset, the design intent was to produce an overtly dramatic building in character with its intended use. Scheme proposals developed from a concept where long roof spans were designed as propped cantilevers either side of a rigid central theatre block. The resulting structural arrangement therefore reflected a classic bending moment diagram for back-to-back propped cantilevers.

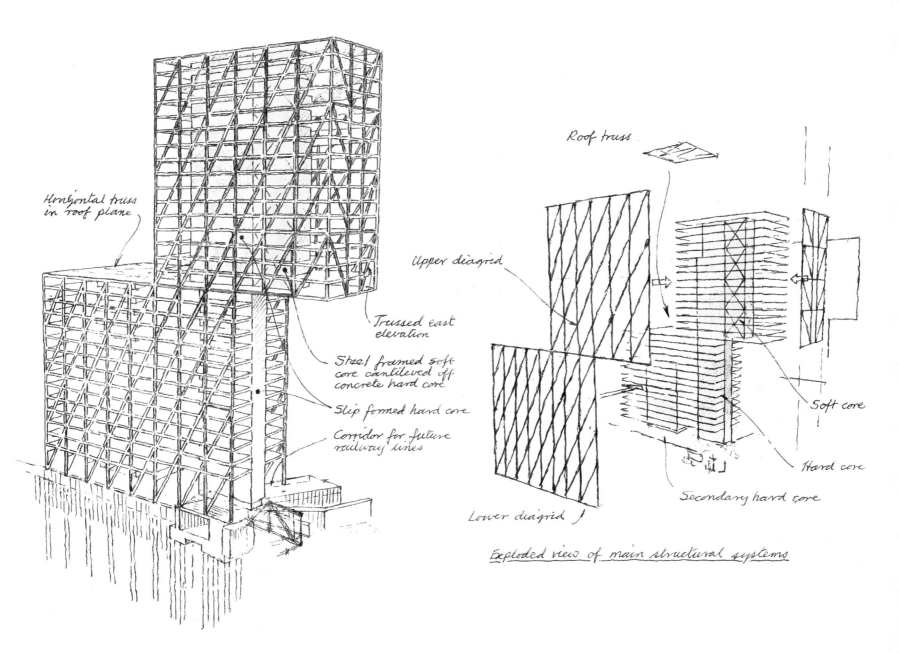

Horizontal truss in roof plane

Trussed east elevation

Steel framed soft core cantilevered off concrete hard core

Slip formed hard core

Corridor for future railway lines

Roof truss

Upper diagrid

Lower diagrid

Soft core

Hard core

Secondary hard core

Exploded view of main structural systems

FIGURE 3.3/12

Cantilever scheme, Northgate, London. This diagrid scheme for the Northgate site cantilevered 27 metres into the air rights space over the railway at the junction of Norton Folgate and Worship Street. It might have qualified as 'extreme engineering' and would have been very difficult to build. The exploded view on the right was used to demonstrate the key structural systems and components. Perhaps surprisingly, control of torsional drift (lateral twisting displacement under wind loading) proved to be the greatest problem.

107

FIGURE 3.3/13

London School of Economics, Bankside, London. The staggered truss system, shown bottom right, is an ingenious way of designing a base structure that provides a number of attractive structural features: reasonably large column-free floor areas, flat soffits, efficient use of steel in storey-high trusses and inherent lateral stability. The central corridor, passing through vierendeel bays, makes the concept particularly appropriate for hotel or student accommodation layouts. At the LSE's Bankside building, the concept was further developed in an over-build scheme which cantilevered at each end above existing structure that was deemed incapable of accepting additional load.

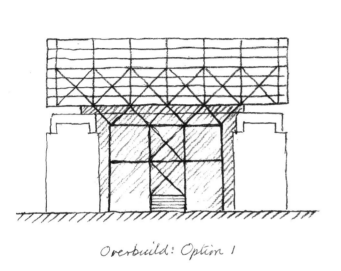

Overbuild: Option 1

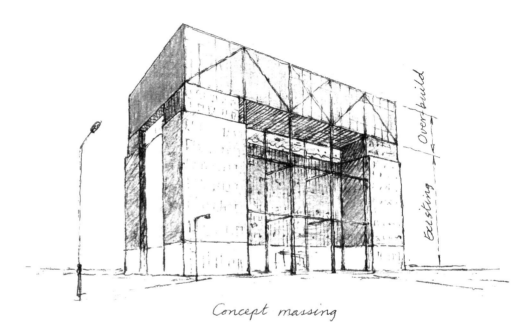

Concept massing

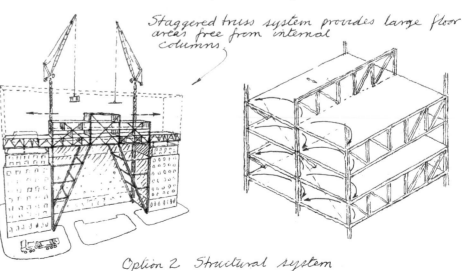

Staggered truss system provides large floor areas free from internal columns

Overbuild: Option 2

Option 2 Structural system

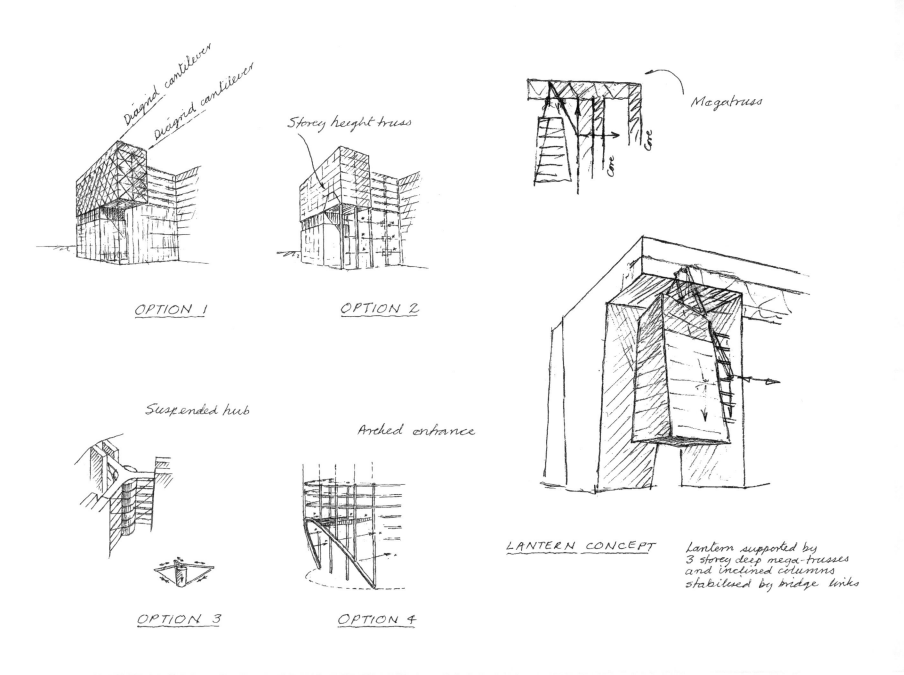

Diagrid cantilever
Diagrid cantilever

OPTION 1

Storey height truss

OPTION 2

Megatruss

Core

Core

Suspended hub

OPTION 3

Arched entrance

OPTION 4

LANTERN CONCEPT

Lantern supported by
3 storey deep mega-trusses
and inclined columns
stabilised by bridge links

FIGURE 3.3/14

China competition. The challenge in some design competitions is to propose something different, something that has not been done before. Here the focus was on the main entrance to a relatively small building that in terms of height and massing presented few opportunities. Ideas ranged from unnecessary cantilevered floors, to giant arched lobbies to a rather bizarre suite of executive offices and meeting rooms suspended from high level designed to resemble a massive Chinese lantern.

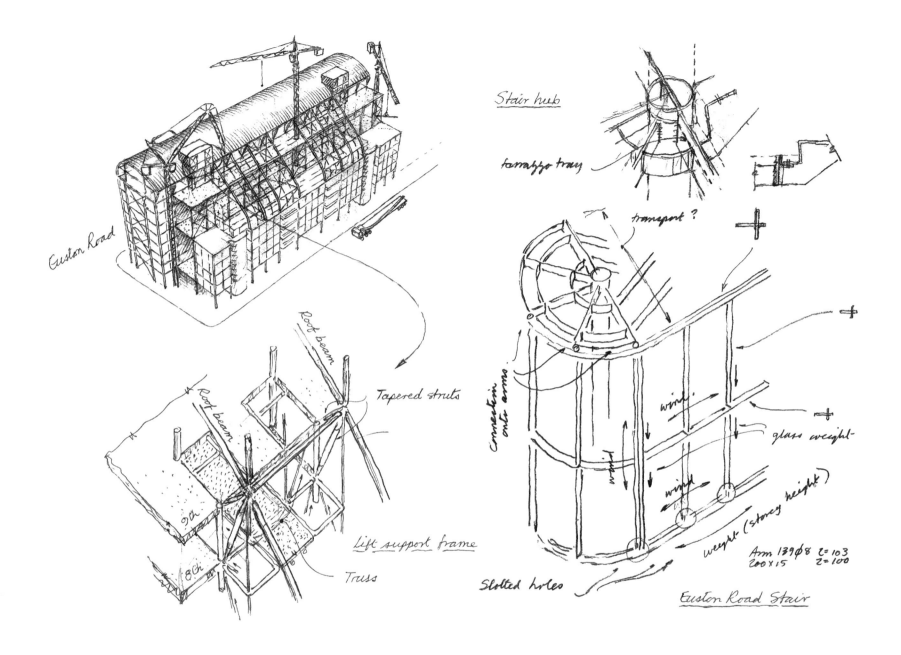

FIGURE 3.3/15

The Wellcome Trust headquarters, Euston Road, London. At the first design team meeting, the architect said it would take five years to design and build a worthy headquarters building for the Wellcome Trust and he wasn't far wrong. The building really consists of two parallel blocks either side of an internal street – inside it is impressive, set off by carefully detailed steel features and high quality finishes. Drawing and developing the expressed structural details brings architect and engineer together with a common aim of producing a building to be proud of.

FIGURE 3.3/16

Building 5, the Shell Centre redevelopment. Before committing to more conventional CAD drawings, RIBA stage C concept design is summarised here using hand drawn sketches of key features. This design is for Building 5 at the Shell redevelopment site on London's South Bank. An axonometric of the more unusual part of the design is intended to add interest and understanding to an otherwise routine structural scheme.

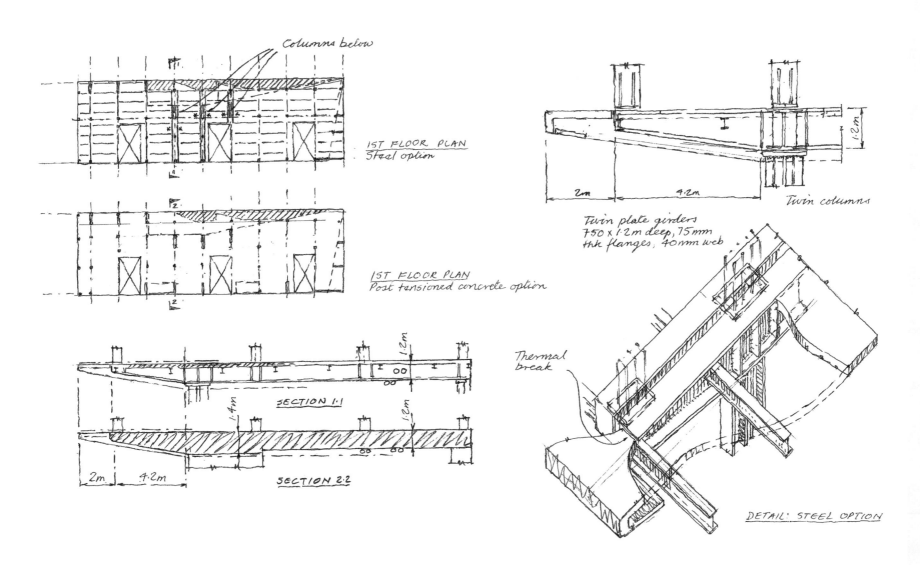

Columns below

1ST FLOOR PLAN
Steel option

1ST FLOOR PLAN
Post tensioned concrete option

SECTION 1.1

SECTION 2.2

2m 4.2m

1.2m
1.4m
1.2m

2m 4.2m

1.2m

Twin columns

Twin plate girders
750 x 1.2m deep, 75mm
thk flanges, 40mm web

Thermal break

DETAIL: STEEL OPTION

SKETCHBOOK

111

Parklands Textiles, Annesley Woodhouse, Nottinghamshire.
Earthmoving is a bit uninspiring but has to be dealt with. The
Parklands Textiles factory in the Midlands was built on spread
footings partly founded on cut areas and partly on compacted
fill material. Sketches were used to explain the strategy and to
impress on the contractors the importance of complying with the
compaction specifications for the filled area of the site.

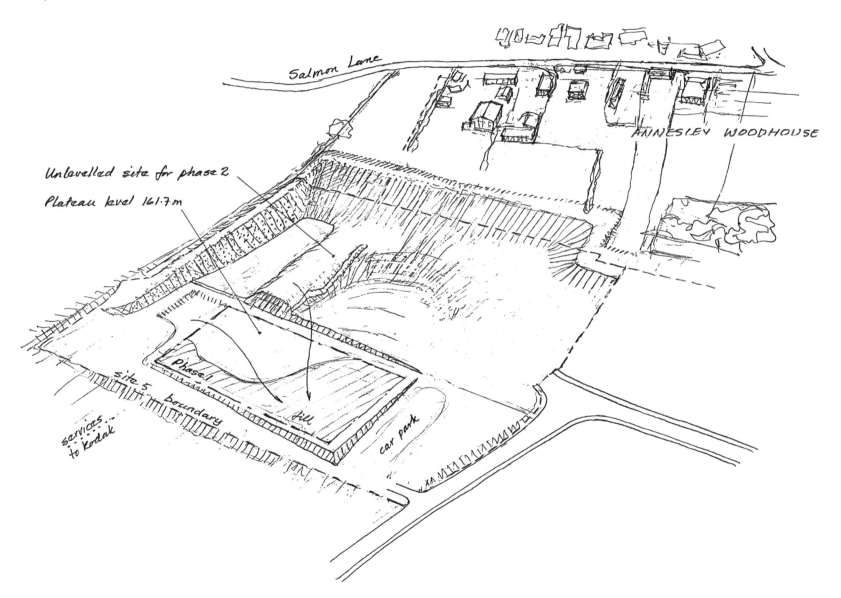

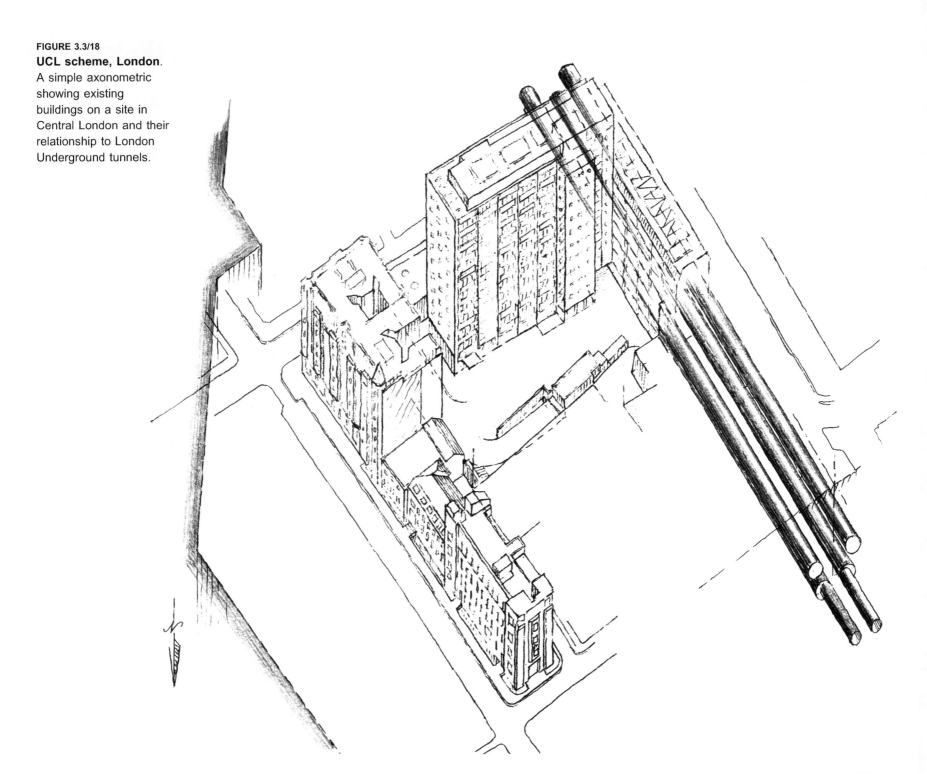

FIGURE 3.3/18
UCL scheme, London.
A simple axonometric
showing existing
buildings on a site in
Central London and their
relationship to London
Underground tunnels.

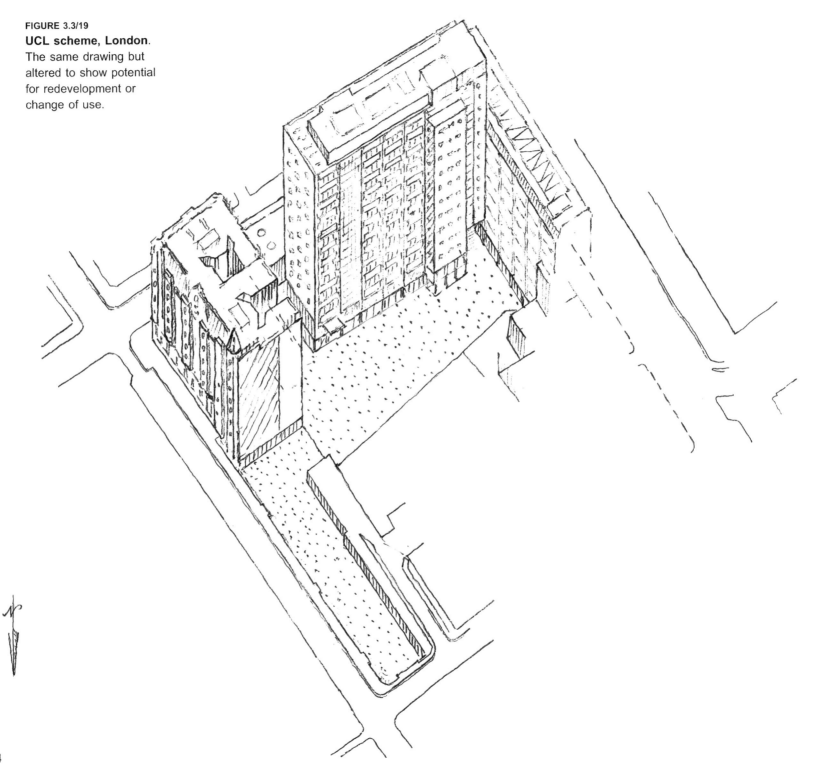

FIGURE 3.3/19
UCL scheme, London.
The same drawing but
altered to show potential
for redevelopment or
change of use.

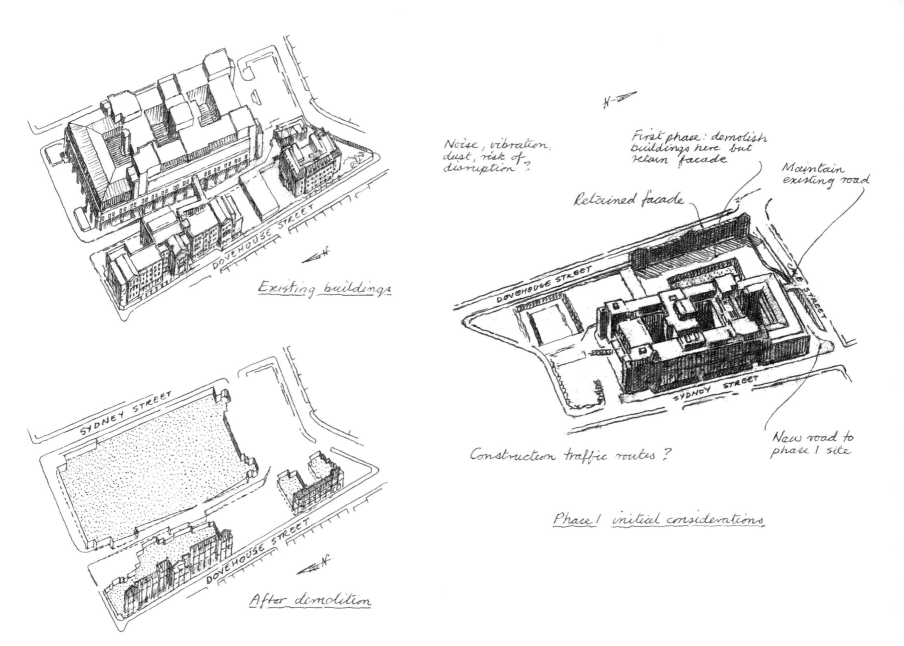

Existing buildings

After demolition

Noise, vibration, dust, risk of disruption?

First phase: demolish buildings here but retain facade

Maintain existing road

Retained facade

Construction traffic routes?

New road to phase I site

Phase I initial considerations

FIGURE 3.3/20

Royal Brompton Hospital, Sydney Street, London. Different schemes were considered for redeveloping the Royal Brompton Hospital's main site in London. Some included demolition of the main building with only the façades of ancillary buildings retained, others looked at relatively minor interventions within the main building, keeping it in use, but totally rebuilding the ancillary buildings behind the retained façades. The second option, of course, would always be problematic in terms of avoiding disruption to the day-to-day running of a busy hospital.

FIGURE 3.3/21

Royal Brompton Hospital, South Parade, London. Drawn using aerial views available on the web and street plans, it is possible to capture a fair amount of detail. However, there are always hidden features and a site visit is essential. Working just with aerial views it is often difficult to judge heights and level changes, and of course aerial views will tell you nothing about existing basements and below ground infrastructure.

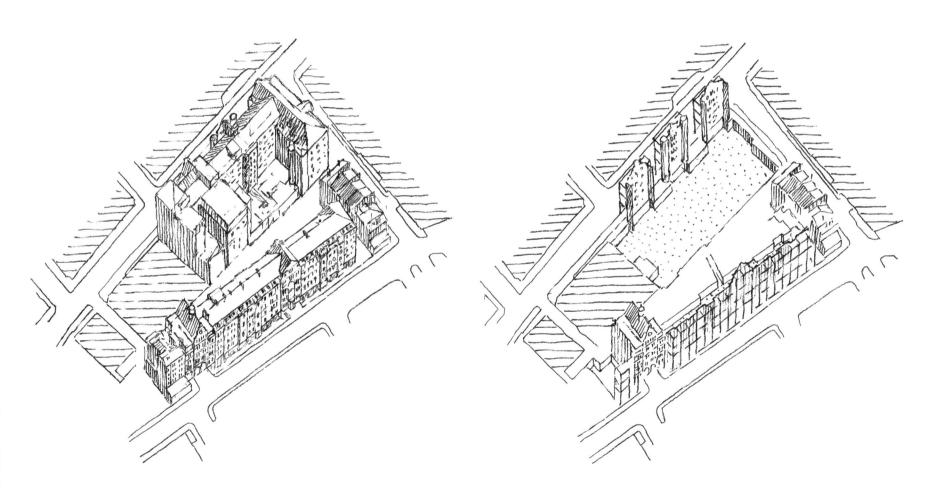

FIGURE 3.3/22

Threadneedle Street, London. Façade design requires input from many disciplines, not least from construction engineers who have to consider transport, logistics and buildability. With the renewed interest in the economic and sustainability credentials of pre-cast concrete, the system illustrated here was proposed for the Threadneedle Street project with Eric Parry Architects but eventually abandoned in favour of a different aesthetic.

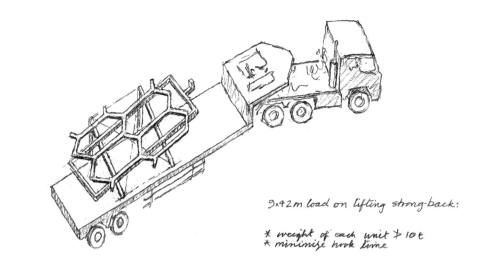

9×4.2m load on lifting strong-back:

* weight of each unit ≯ 10 t
* minimise hook time

LOGISTICS

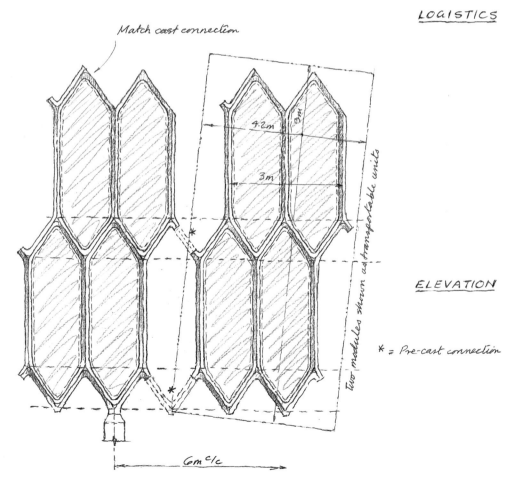

Match cast connection

4.2m

3m

6m c/c

Two modules shown as transportable units

ELEVATION

* = Pre-cast connection

3.4   BRIDGES

Bridge design is recognised as a specialised and rather glamorous area within structural and civil engineering.

Over the years there has been much debate about the role of the various professions in the design of a bridge. Many believe that bridge design should be led by the engineer and that the architect should play a secondary, beautifying role. Some believe a bridge designed to be efficient and economic will have an inherent aesthetic quality without input from architects at all, and others believe engineers need help from architects to choose and develop an idea from a range of possibilities.

In a way concept design is straightforward because, some would argue, there are only seven basic bridge forms: the beam, the truss, the arch, the suspension, the cantilever, the portal frame and the cable-stay; but of course basic forms can be combined and hybrids can be developed. Spans can be lifted or retracted to allow the passage of ships, for instance, and different materials can be used leading to yet more options or sub-types.

Bridge ideas in my experience are best developed beyond concept stage with engineer and architect working side by side, but the responsibility for knowing what is achievable and practical must lie firmly with the engineer. There is no doubt that the skill of specialist design engineers is essential. This is because load patterns on a bridge deck are complex, wind and seismic loadings are critically important and dynamic behaviour and other factors such as design for movement, safety and economics must be taken into account. Fatigue is a factor, particularly for road and rail bridges where engineers spend a lot of time looking not only

at pattern loads but at traction, braking and lateral forces. In addition, the buildability of a bridge, often in difficult locations across chasms, roads, railways or water, should influence which design is to be adopted. Long term maintenance, the replacement of bearings and protection against the elements must all be given due consideration.

The leading bridge design codes of practice are complex. For instance, the current edition of the Eurocodes covering bridge design runs to over 530 pages, has 38 sections, 165 sub-sections and 31 appendices. The complexity inherent in bridge design and analysis has led to the development of specialist design techniques, advanced modelling and complex analytical software.

Concept sketching of bridge design is often especially rewarding because the basic forms are simple and easy to portray. There is also the question of the build sequence to consider, and perhaps there is a need to look at moving elements of a swing or lifting bridge. These all create a rich area for exciting and interesting sets of drawings.

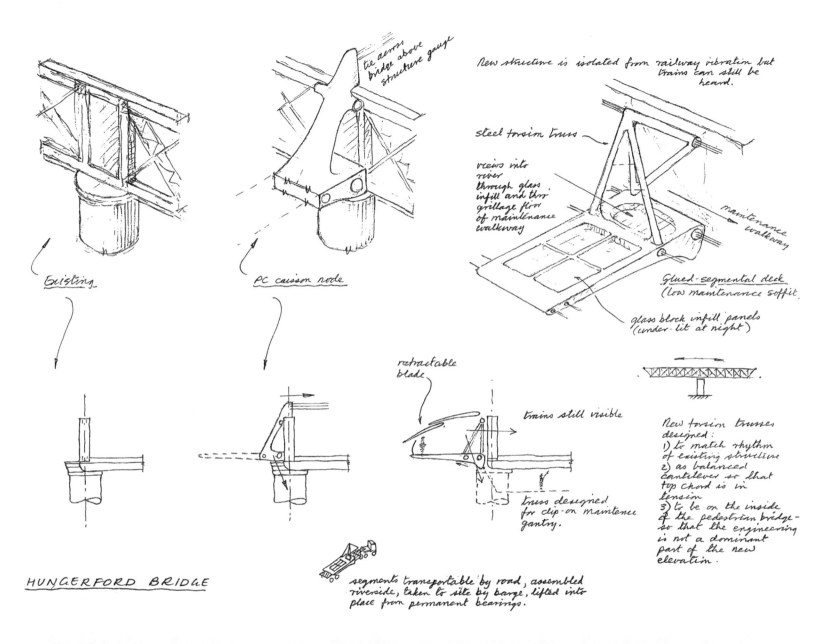

tie across bridge above structure gauge

New structure is isolated from railway vibration but trains can still be heard.

steel torsion truss

views into river through glass infill and thin grillage floor of maintenance walkway

maintenance walkway

Existing

PC caisson node

Glued-segmental deck (low maintenance soffit

glass block infill panels (under-lit at night)

retractable blade

trains still visible

truss designed for clip-on maintence gantry.

New torsion trusses designed:
1) to match rhythm of existing structure
2) as balanced cantilever so that top chord is in tension
3) to be on the inside of the pedestrian bridge- so that the engineering is not a dominant part of the new elevation.

HUNGERFORD BRIDGE

segments transportable by road, assembled riverside, taken to site by barge, lifted into place from permanent bearings.

FIGURE 3.4/1

Hungerford Bridge design competition. The Hungerford Bridge design competition was organised to improve cross-river connectivity and to replace dilapidated and inadequate walkways on both sides of the Charing Cross railway bridge. This bridge, by Sir John Hawkshaw, for the South Eastern Railway, replaced an earlier suspension footbridge by I. K. Brunel. The footbridge chains were removed for use on the Clifton Suspension Bridge, Bristol. Hawkshaw's rather industrial design consists of nine wrought-iron trusses supported on cast-iron cylinders and on the two arched brick river-piers constructed for Brunel's suspension bridge. Working with architect Lifschutz Davidson Sandilands, our early designs were based on clip-on trusses designed to match the rhythm of the Hawkshaw trusses.

FIGURE 3.4/2
Hungerford Bridge design competition. Later competition designs moved away from references to the Hawkshaw trusses towards designs based on pylons, a theme used extensively in the 1951 Festival of Britain for the 'Skylon' and flag staff features.

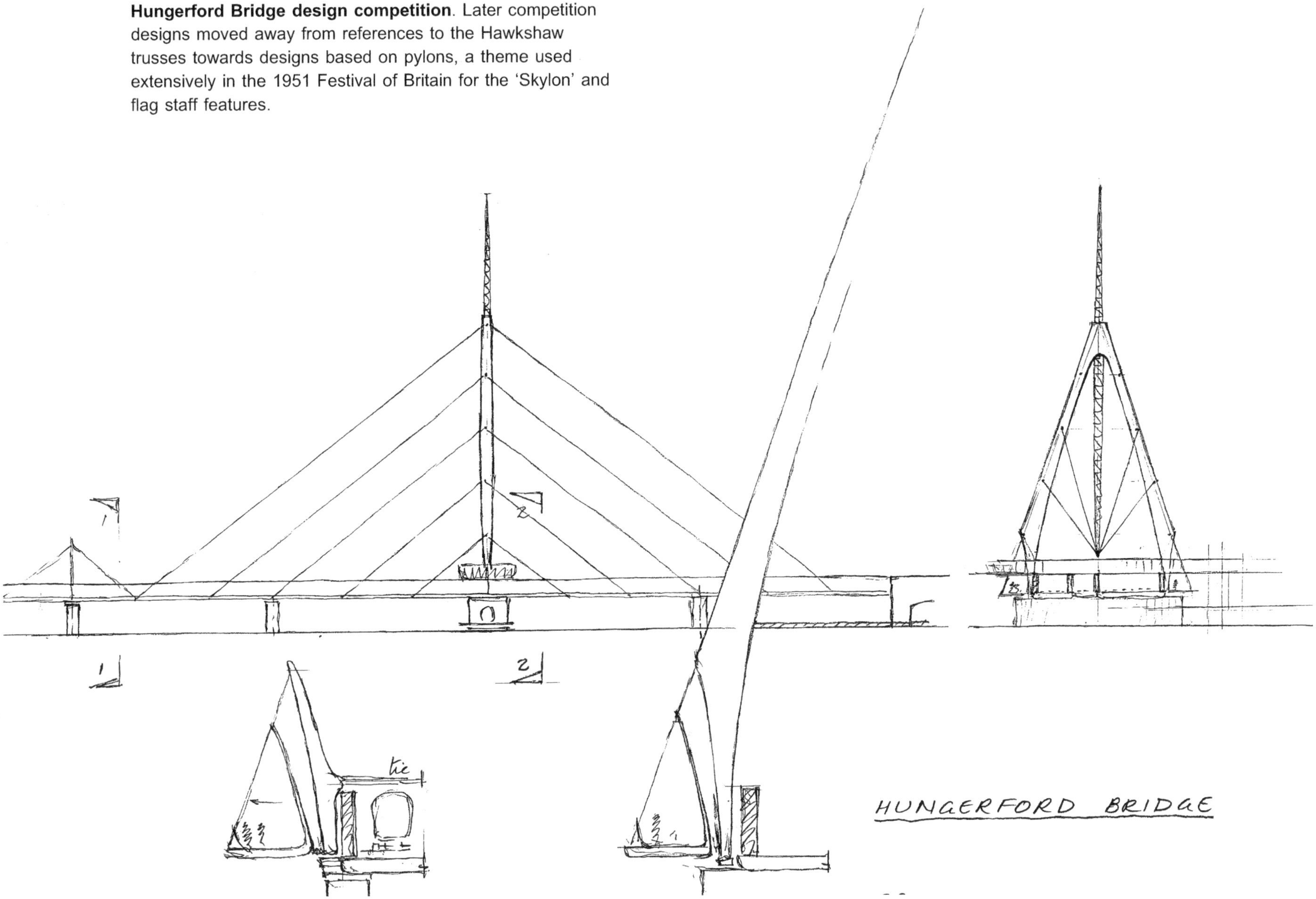

HUNGERFORD BRIDGE

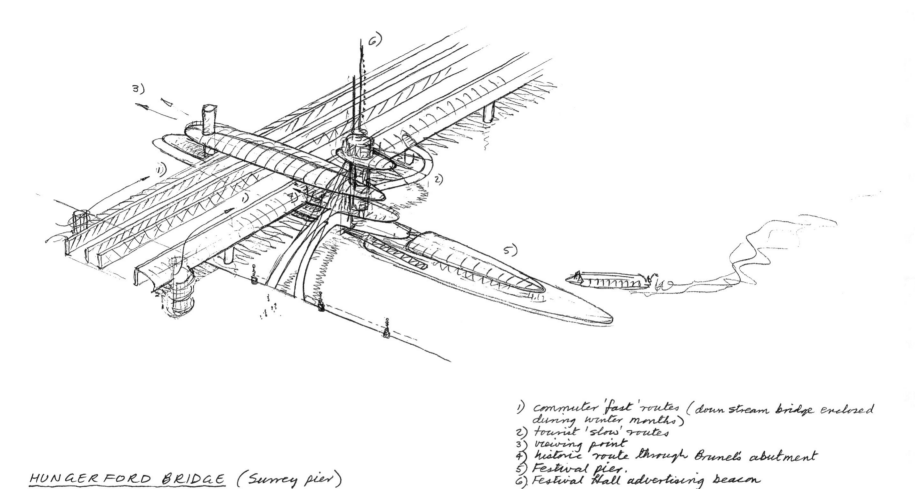

HUNGERFORD BRIDGE (Surrey pier)

1) commuter 'fast' routes (down stream bridge enclosed during winter months)
2) tourist 'slow' routes
3) viewing point
4) historic route through Brunel's abutment
5) Festival pier.
6) Festival Hall advertising beacon

FIGURE 3.4/3

Hungerford Bridge design competition. More design studies concentrated on the southern Brunel pier which still stands in the river (since the construction of the Victoria Embankment by Sir Joseph Bazalgette, the northern pier is no longer surrounded by water). Brunel's southern pier was seen as a location that could be exploited as a viewing point for tourists making use of a convenient existing passageway through the pier from one side of the railway bridge to the other. An idea to use secondary diagonal foot bridges connecting the South Bank to the Brunel pier emerged to promote the concept of a circular tourist walk from Parliament, over Westminster Bridge, along the South Bank, across a diagonal bridge onto the upstream main pedestrian bridge to Victoria Embankment and then back to the Houses of Parliament.

FIGURE 3.4/4

Hungerford Bridge design competition. Combinations and variants of ideas were tested: pylons, covered walkways; one walkway only covered for 'winter use'; the tourist route and a floating extension to the pier acting as a landing stage for people using river boats – all these ideas were examined and evaluated. Cost, however, had to be kept in check, so some ideas were abandoned and others, like the diagonal secondary bridges, left as a possibility for future generations. More fundamentally, there was no logical justification for a super-tall set of pylons centred on the southern pier and so this particular element was removed from the design.

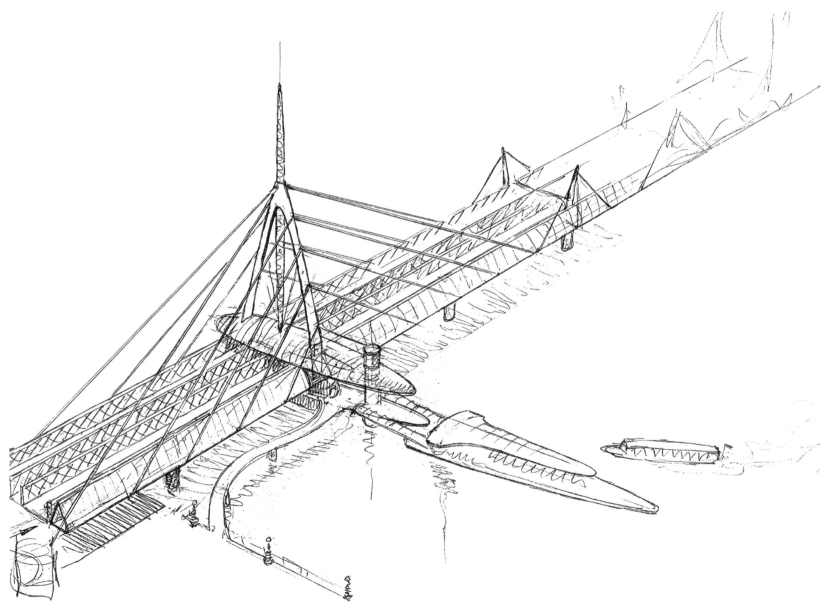

FIGURE 3.4/5

The Golden Jubilee Bridges, Hungerford Bridge, London. Several erection methods were studied in an effort to find the most practical and economic construction strategy. This version was based on pre-cast sections of the bridge being brought to site by barge and then lifted onto temporary works towers. At the time the bridge decks were tapered (on plan) with the widest part in the middle of the river. An architectural change led to a constant width walkway which saved cost, and this in turn led to a methodology change whereby the deck was cast on shore and continuously jacked from one bank to the other.

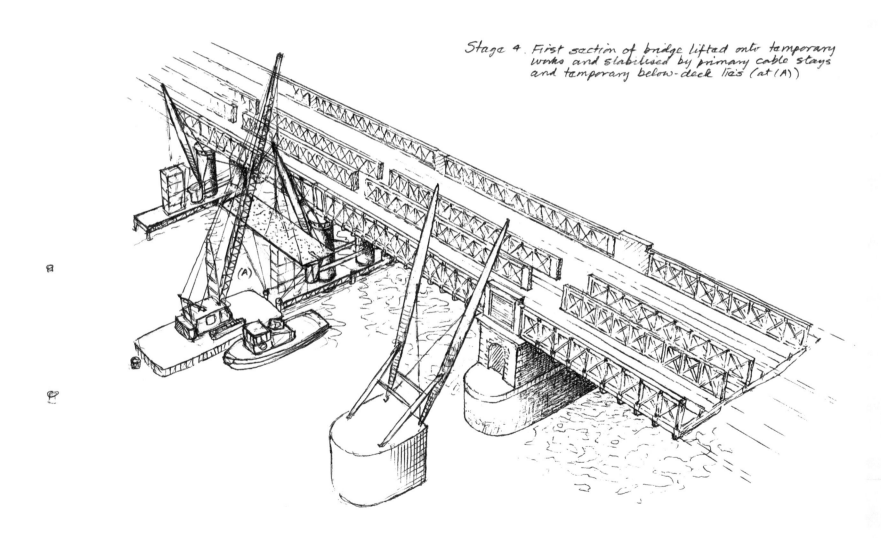

Stage 4. First section of bridge lifted onto temporary works and stabilised by primary cable stays and temporary below-deck ties (at (A))

SKETCHBOOK

FIGURE 3.4/6
The Golden Jubilee Bridges, Hungerford Bridge, London. The original competition schemes were based on designs where the new pedestrian bridges (one either side of Hawkshaw's bridge) were supported off the railway bridge itself, rather like the old walkways had been. But it soon became apparent that the existing caissons were not in good condition and had become quite vulnerable to ship impact. The outcome was a design that did not rely on existing structure and that had new foundations in the river, comprising cutwater structures that would in fact protect the caissons. The sketch is not so different to the as-built structures.

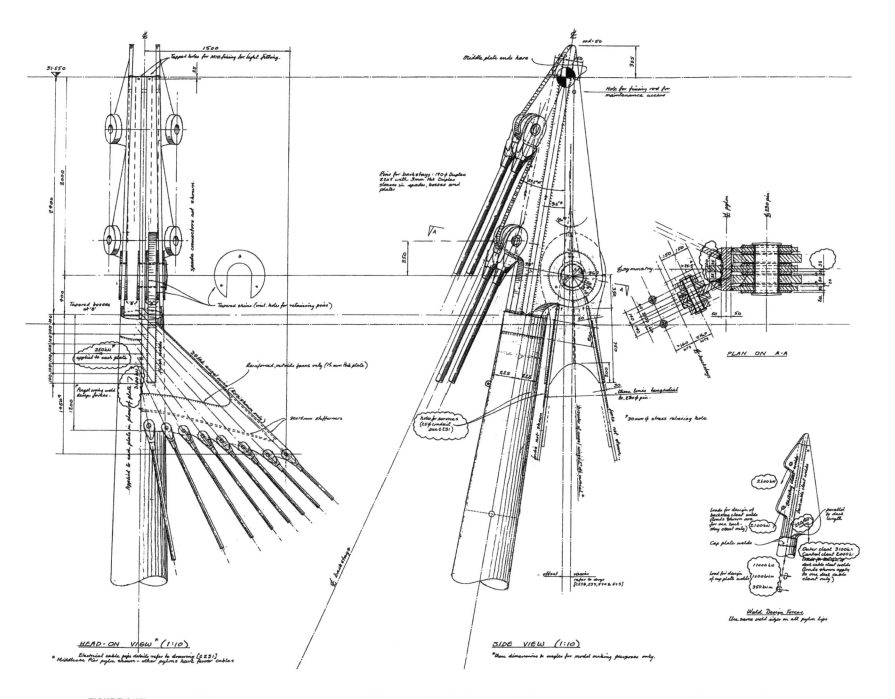

FIGURE 3.4/7

The Golden Jubilee Bridges, Hungerford Bridge, London. This is the detail at the top of the Hungerford Bridges' pylons –
although a conventional, if old fashioned, scale drawing and not strictly speaking a hand sketch, it was based on set out dimensions
and concept sketches. Just getting various components to fit together in an organised way can be very challenging.

125

FIGURE 3.4/8

Waterloo East Link Bridge, Waterloo Station, London. The high level pedestrian link bridge between Waterloo Main Station and Waterloo East was constructed over the route of an early rail link that once crossed Waterloo Road. The old rail bridge is still in place and was used for temporary support during the erection of the new structure. A photograph taken from the top of a tall building adjacent to site was used as the background for this sketch.

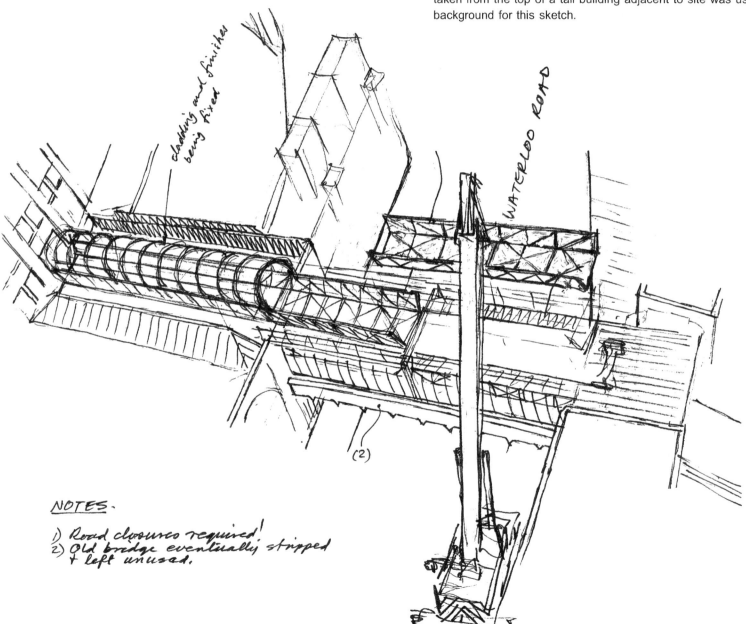

cladding and finishes being fixed

WATERLOO ROAD

(2)

NOTES.
1) Road closures required!
2) Old bridge eventually stripped + left unused.

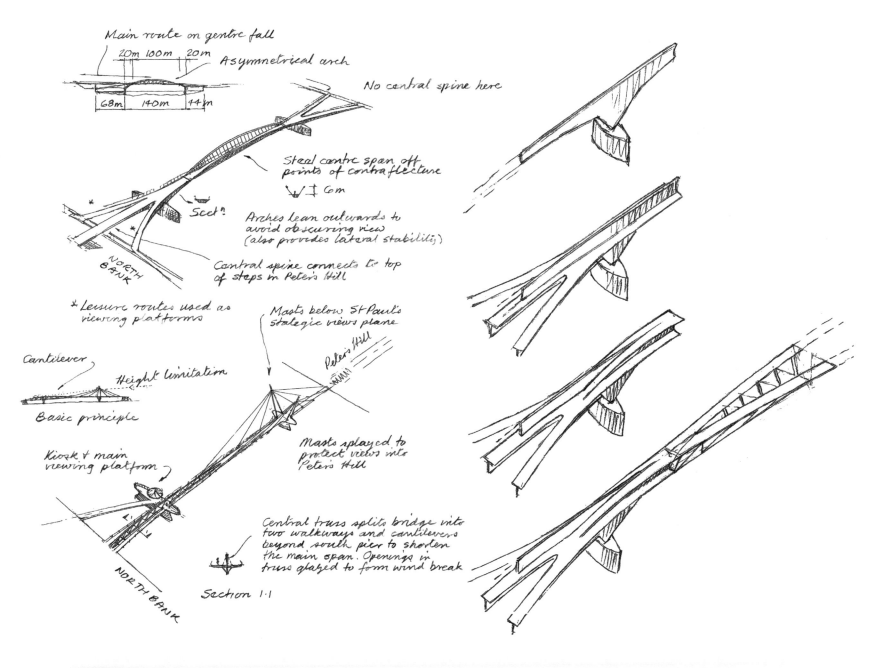

Main route on gentre fall

20m 100m 20m Asymmetrical arch

68m 140m 44m

No central spine here

Steel centre span off points of contra flecture

⌄ ‖ 6m

Arches lean outwards to avoid obscuring view (also provides lateral stability)

Central spine connects to top of steps in Peter's Hill

Sect⁰

NORTH BANK

* Leisure routes used as viewing platforms

Masts below St Paul's stategic views plane

Peter's Hill

Cantilever

Height limitation

Basic principle

Masts splayed to protect views into Peter's Hill

Kiosk + main viewing platform

Central truss splits bridge into two walkways and cantilevers beyond south pier to shorten the main span. Openings in truss glazed to form wind break

NORTH BANK

Section 1·1

SKETCHBOOK

FIGURE 3.4/9

Millennium Bridge design competition. Another design completion with Chetwood Architects, this time for what is now the Millennium Bridge between St Paul's Cathedral on the north bank of the Thames and Tate Modern on the South Bank. Unlike the winning design, we took the pragmatic approach of providing the minimum acceptable central span and used the side spans to assist the middle span either by cantilever action or by a cable stay system. Splayed structure is used to maintain views of St Paul's from the south and the inclination of St Paul's sight lines generate an interesting degree of asymmetry.

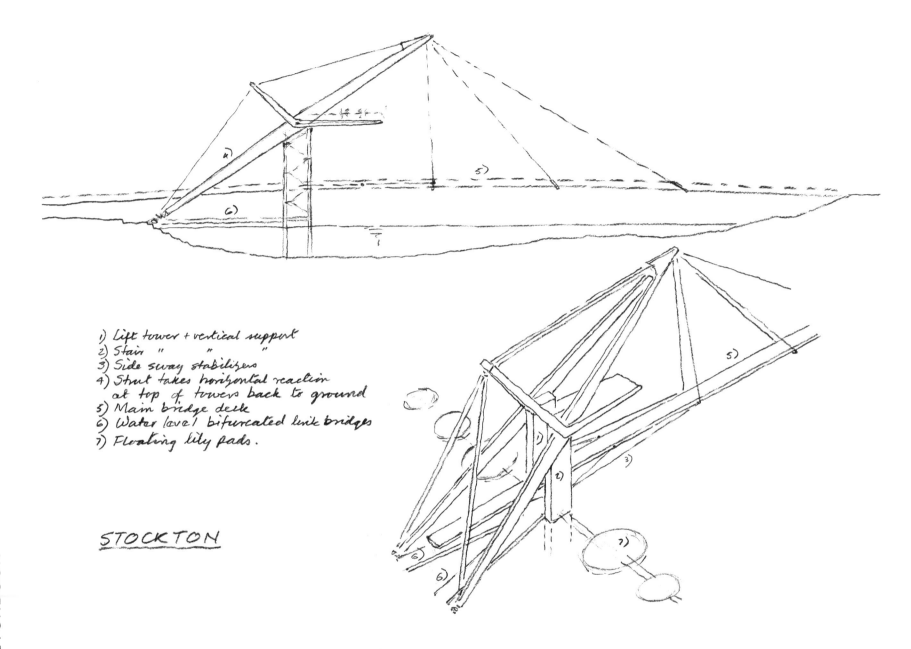

1) Lift tower + vertical support
2) Stair   "   "   "
3) Side sway stabilizers
4) Strut takes horizontal reaction
   at top of towers back to ground
5) Main bridge deck
6) Water level bifurcated link bridges
7) Floating lily pads.

STOCKTON

FIGURE 3.4/10
Stockton Pedestrian Bridge competition. This design was developed with Lifschutz Davidson Sandilands and was inspired by the architecture of industry and by the form of cranes in particular. The concept was to use muscular primary structure consisting of twin cantilever arms but softened aesthetically by slender cables, sway stabilizers and even floating appendages. Getting the proportioning and spatial arrangement of the various components correct was important both visually and from an engineering point of view. Poorly proportioned arrangements lead to inefficiency. For example, if the angle between the cable and the deck is too acute, then the cables are ineffective.

Stockton Pedestrian Bridge competition – erection sequence. (1) Marine piling, (2) Cofferdam construction for the foundation of the main piers, (3) Slip forming the piers. The piers support the cantilevers but also house stairs and lifts. (4) Erection of strand-jack gantry, (5) Cantilever arms assembled on the construction platform, (6) Cantilever arms lifted into place using strand-jack gantry.

Stockton Pedestrian Bridge competition – erection sequence. (7) Single cantilever arm in place, (8) Second cantilever arm in place, (9) First bridge span lifted into position from barge, (10) Second bridge span lifted into position from barge, (11) Third bridge span floated into position.

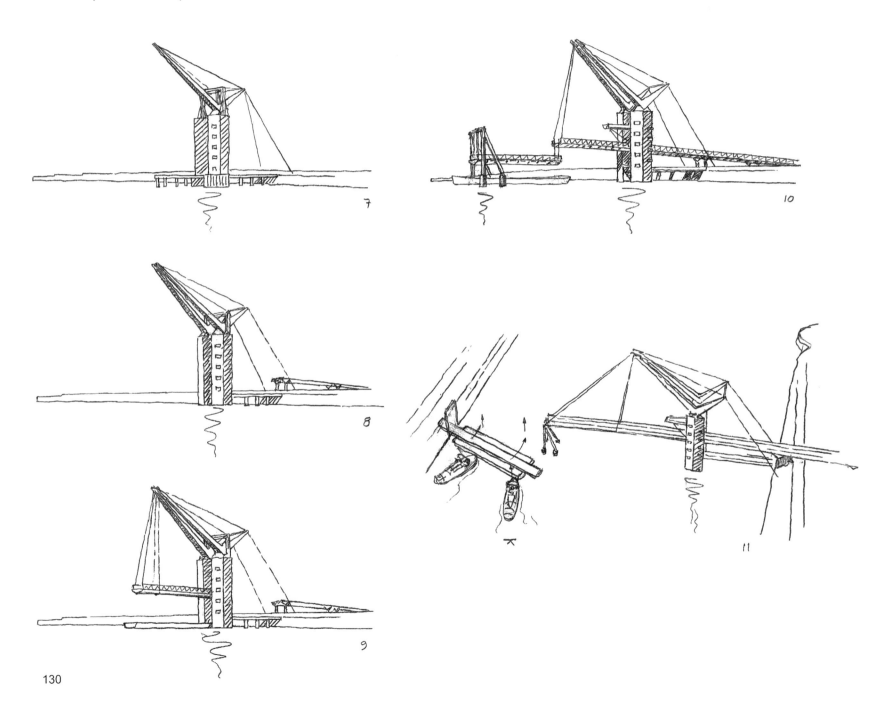

SKETCHBOOK

Pedestrian bridge, University of Northumbria, Newcastle. Sometimes the first simple ideas are the best. This pedestrian bridge connecting different parts of the university campus changed little from the scheme shown on the initial concept sketch to the built structure. The deck was constructed on the bank parallel to the road and rotated into position. (The secondary span to the right lost its pylon and is now just a beam spanning a railway.)

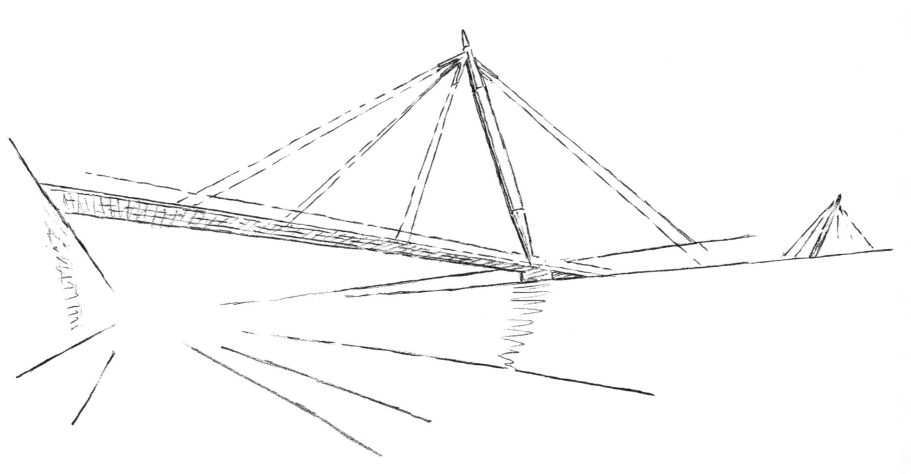

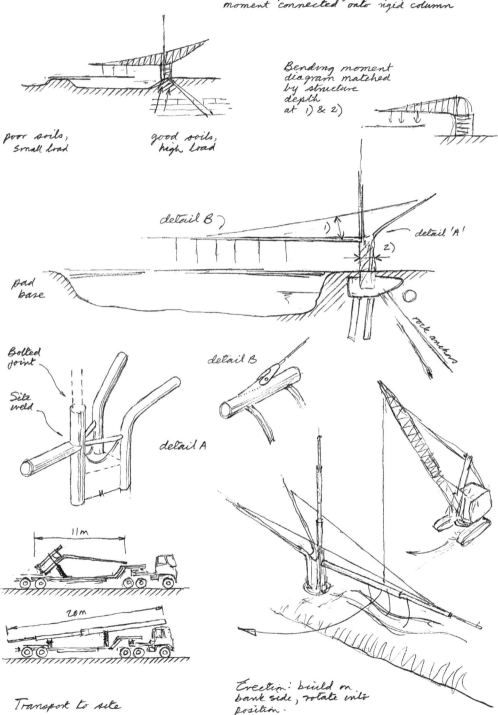

BEDFORD

Main support on north bank – cable-stayed boom forms deep beam moment connected onto rigid column

poor soils, small load

good soils, high load

Bending moment diagram matched by structure depth at 1) & 2)

detail B)

detail 'A'

pad base

rock anchors

Bolted joint

Site weld

detail A

detail B

11 m

20 m

Transport to site

Erection: build on bank side, rotate into position.

FIGURE 3.4/14
Bedford Pedestrian Bridge design competition. Inspiration doesn't often come from the competency of the ground strata, but here different rock on either side of a small river led to the adoption of an asymmetrical solution. The resulting rather modest design, capable of full off-site prefabrication, was developed with Chetwood Architects and the design was eventually runner-up to the competition winner, Santiago Calatrava. The left-hand bank supports a minor load, while the right-hand structure and its foundation carries by far the greater part of the weight of the bridge.

SKETCHBOOK

132

3.5   AIRPORTS

Airports are emotional places, for some people the gateway to
exciting parts of the world, for others not only points of arrival and
encounter but also of departure and separation. They certainly get
a lot of attention from all sectors of the community and are so
often the focus of national pride and a showcase for futuristic
design. International terminal buildings tend to be on a
monumental scale, where architects and planners attempt to
develop a sense of grandeur. Meeting spatial demands and the
creation of a sense of space are crucial objectives, and of course
designers turn to ethereal concepts such as lightness and
movement.

However, we still seem to be grappling with the challenges of
airport design – passenger flows, security, baggage handling,
infrastructure connections, retail and myriad other elements, but
also the environmental impact of these 'cities within cities'. In a
way, structural engineering may seem very secondary to these
wider and very important issues.

There is of course, the less glamorous side of airport design:
control towers, hangars, cargo facilities, car parks, extensions to
existing facilities, all bring their special requirements.

The HET Project at Heathrow Airport, London. Rebuilding a major terminus at one of the world's busiest airports requires intense and detailed planning. Master-planning is really the first step and it relies on high level understanding of existing buildings, existing infrastructure and how the terminus operates within the airport as a whole. The main sketch provides a broad picture of one of the important constraints, underground infrastructure, and the construction sequence diagrams on the right are from a report showing how the new building might be built to replace the existing facilities incrementally while keeping the terminus operational.

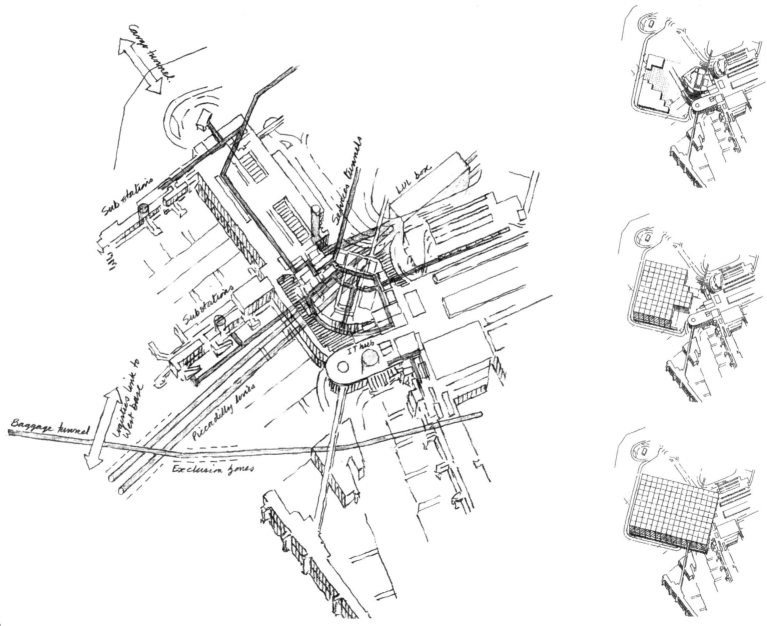

FIGURE 3.5/2

Generic airport, USA. Master-planning for an idealised generic provincial airport could only be meaningful in North America or possibly the Far East. The idea was to develop a layout that could be expanded by adding satellites and then mirrored to cope with increased passenger flows when a second runway became necessary. Key factors were orientation to suit prevailing wind directions, existence of local rail and road infrastructure and planning to always allow uninhibited access for construction traffic during a construction process that could take many years.

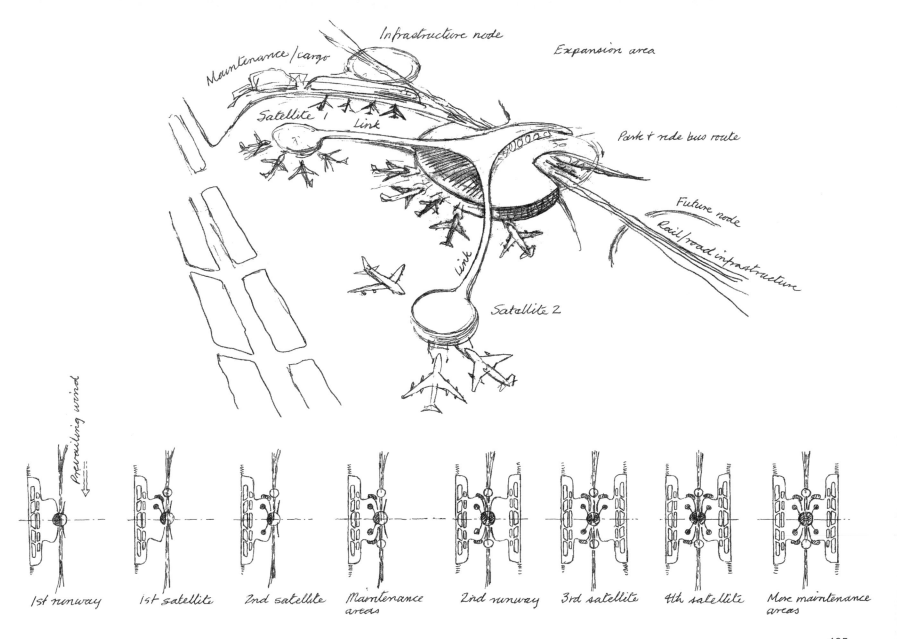

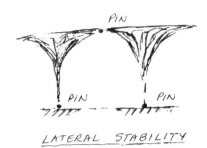

LATERAL STABILITY

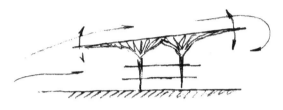

DYNAMIC BEHAVIOUR

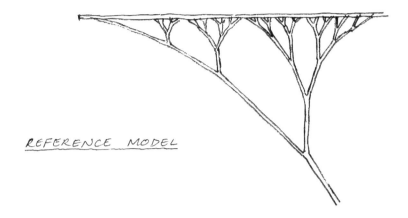

REFERENCE MODEL

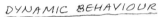

CONSTRUCTION SEQUENCE

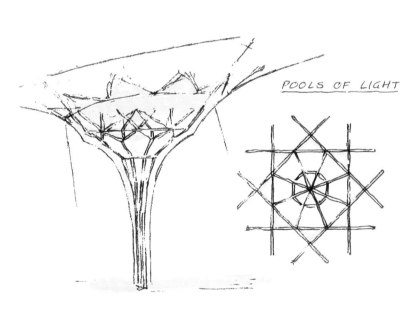

POOLS OF LIGHT

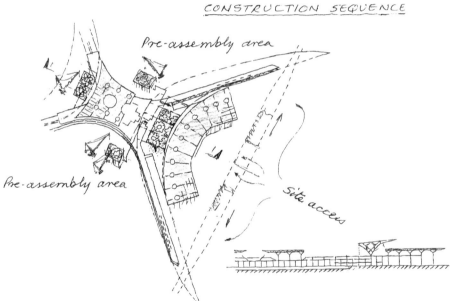

Pre-assembly area

Pre-assembly area

Site access

FIGURE 3.5/3

Airport scheme, Middle East. Trees and tree-like structures are often the inspiration for airport roofs. Here the challenge was to develop a proposal based on the concept sketch (top right) into a viable tree element that could be replicated and interconnected to provide lateral stability and long perimeter cantilevers.

Airport scheme, Middle East. Only mobile cranes could be used here to erect the roof structure of an extension to a Middle Eastern airport because of height and working time restrictions. The roof structure is a cable stiffened beam designed for wind reversal loads and portalised with perimeter columns. The beam and columns work together to form a frame (or portal) capable of resisting gravity and lateral loads. The roof panel sections are assembled at ground level and lifted by cranes working in tandem during engineering hours.

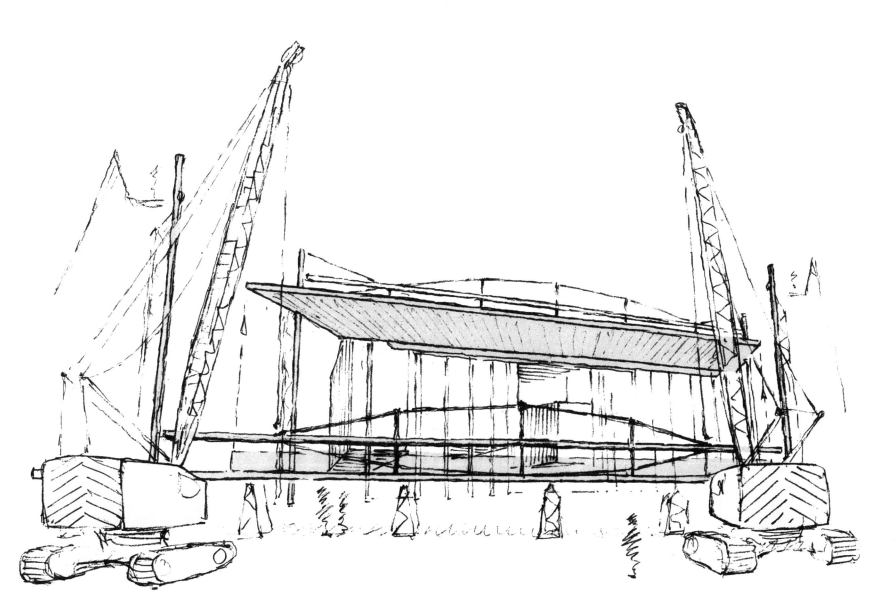

SKETCHBOOK

Airport scheme, Middle East. The design progression moves from fixed bearings for the arches, to interconnecting primary arches, secondary catenary beams and tertiary anti-clastic cable nets. These are quick sketches to initiate the design process which starts with a concept tested by simple hand calculation, possible even on a structure as complex as this, but followed rapidly by more rigorous analysis.

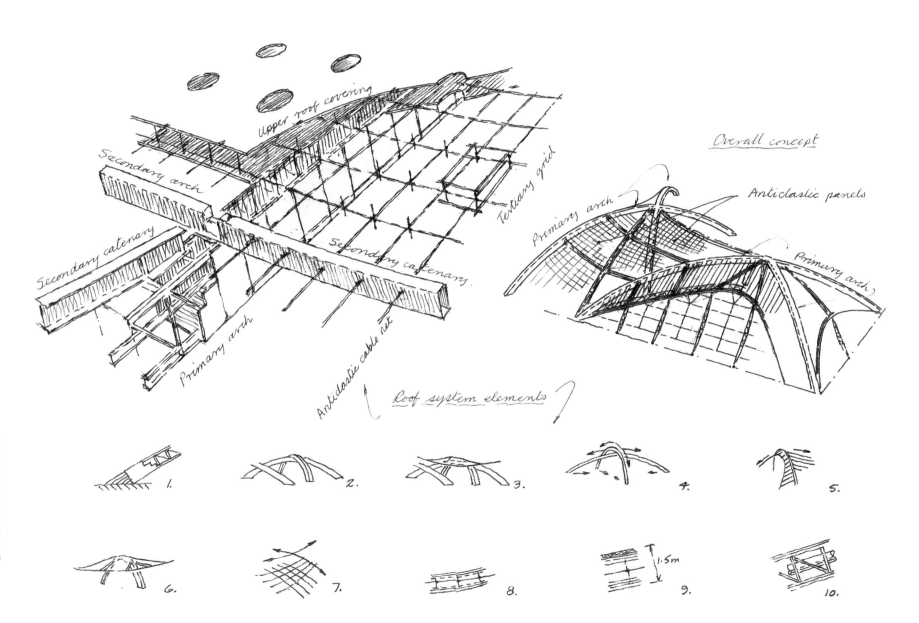

SKETCHBOOK

FIGURE 3.5/6
Northern European airport extension. High level walkways and people-movers give lots of opportunity for combined architectural and structural expression. Here the approach was to find a dynamic two-dimensional solution at the main column positions and then to look at the linear nature of the walkway. The idea developed into a three-dimensional composition and not surprisingly a tree form emerged.

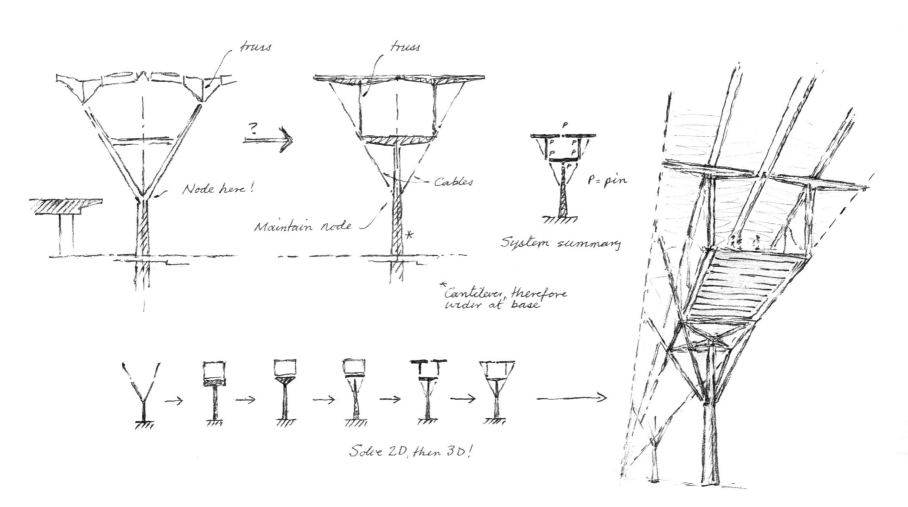

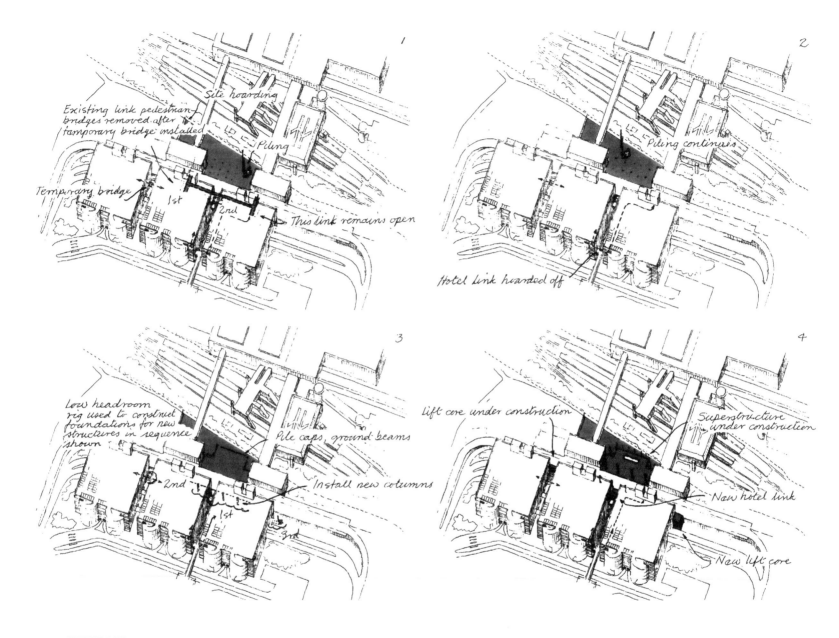

Within the image, handwritten annotations read:

Panel 1
- Existing link pedestrian bridges removed after temporary bridge installed
- Site hoarding
- Piling
- Temporary bridge
- 1st
- 2nd
- This link remains open

Panel 2
- Piling continues
- Hotel link hoarded off

Panel 3
- Low headroom rig used to construct foundations for new structures in sequence shown
- Pile caps, ground beams
- 2nd
- 1st
- 3rd
- Install new columns

Panel 4
- Lift core under construction
- Superstructure under construction
- New hotel link
- New lift core

FIGURE 3.5/7

Car park, South Terminal, London Gatwick Airport, West Sussex. Construction within transport hubs is always challenging. Re-organising Gatwick's South Terminal car park and building between the car park and the railway needed careful planning to keep disruption of vehicle and people movement to a minimum. Four of sixteen phasing diagrams, all based on a common background, are shown here and were developed with the benefit of input from contractor planning experts.

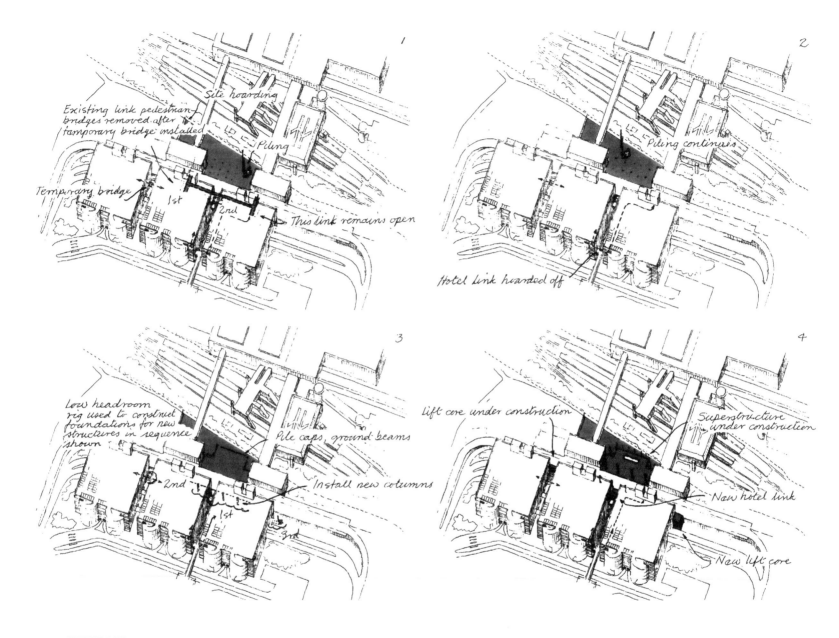

SKETCHBOOK

140

3.6 RAILWAY STATIONS AND BUILD-OVER PROJECTS

Any construction work near or over operating railways or roads, or alterations to stations or other associated infrastructure buildings, is going to be tough. Quite rightly, the top priority for the regulatory authorities is to run a transport system and to run it safely. Not surprisingly therefore, design engineers are often faced with complex approval processes which can sometimes seem highly bureaucratic. This includes the often lengthy 'Approval in Principle' (AIP) or Conceptual Design Statement (CDS) process, where the whole design approach and erection methodology strategy has to be spelt out in unambiguous detail.

In addition, there are unusual physical challenges to deal with, emanating from such things as:

- Very high impact loads (from derailments or errant road vehicles).
- Complex rolling load patterns (axle loads in specific arrangement travelling across a structure).
- Long spans.
- Specific dimensional constraints including dynamic envelopes. (Rail vehicles have very tightly controlled envelopes dependent on dynamic behaviour, track curvature and so on.)
- Sight lines.
- Asset protection. (The regulatory authority's duty is to ensure their infrastructure is not adversely affected by construction works.)
- Acoustic and vibration isolation.
- Maintenance (bridge bearings for instance).

- Crash decks (protection from damage by falling objects during the course of construction).
- Safety, especially with regard to piling or craneage beside the railway or road. (Very definite rules exist about how and when it can be carried out.)
- Relatively unusual structures such as embankments, cuttings, masonry arches, over-bridges and under-bridges.
- Launching, jacking and sliding techniques (to deal with physical constraints beside operational roads or railways).
- Restricted times when construction activities can be carried out (often limited to 'engineering hours' in the middle of the night when the transport is shut down).
- Ground movement (all new construction has the potential to cause ground movement and this of course can affect assets).

On the positive side, existing structure is often well built and robust and in the UK at least, there's a good chance of finding records of existing structures although they may not have the status of 'as-built' drawings.

Planning, programming, clarity and build sequence are especially important, and so is the need to communicate with technical and non-technical people at different levels. Clear sketches and drawings greatly enhance text or narrative.

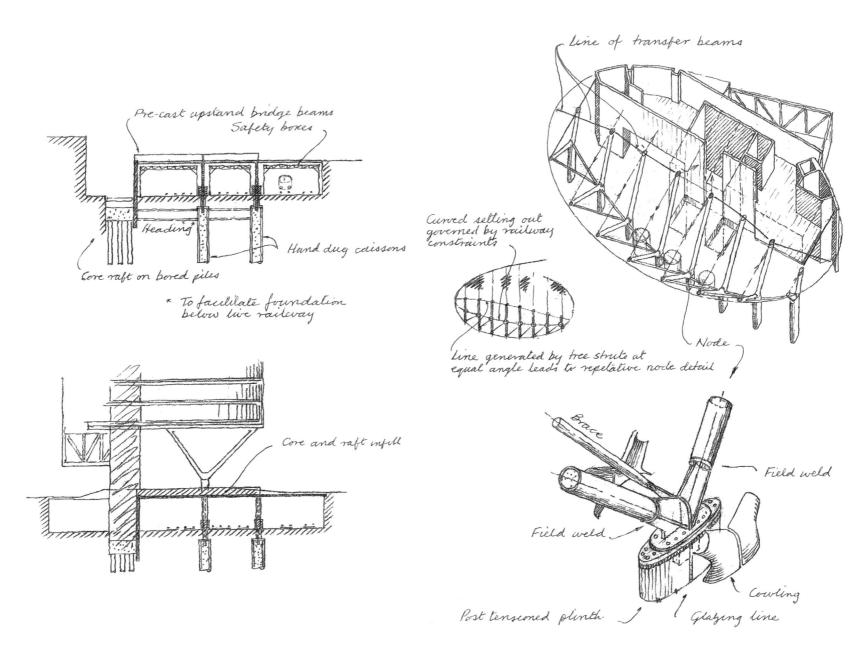

Pre-cast upstand bridge beams
Safety boxes

Heading*

Core raft on bored piles

Hand dug caissons

* To facilitate foundation below live railway

Core and raft infill

line of transfer beams

Curved setting out governed by railway constraints

Line generated by tree struts at equal angle leads to repetative node detail

Node

Brace

Field weld

Field weld

Cowling

Post tensioned plinth

Glazing line

FIGURE 3.6/1
Site at Norton Folgate, London. These sketches represent our earliest involvement with this highly constrained site at Norton Folgate just north of Liverpool Street Station in London. The KPF scheme for a 'build over' project was located above six busy railway lines running through an existing cutting. Space between the tracks for support and foundations was extremely restricted and the railway downtime virtually non-existent. This led to the exploration of unusual construction techniques based on headings for access and foundations consisting of hand dug caissons.

FIGURE 3.6/2

Deck over railway cutting, Norton Folgate, London. Decking over the railway cutting at Norton Folgate proved to be a very difficult design challenge due to the many site constraints. Depth of new structure was limited on the underside by rail requirements and on the top surface by sight lines to the new development. Locations for new foundations were virtually non-existent and building over an operating railway is never easy. The undulating surface in the right hand sketches owed much to the architect's recently completed extension to the Smithsonian Museum in Washington.

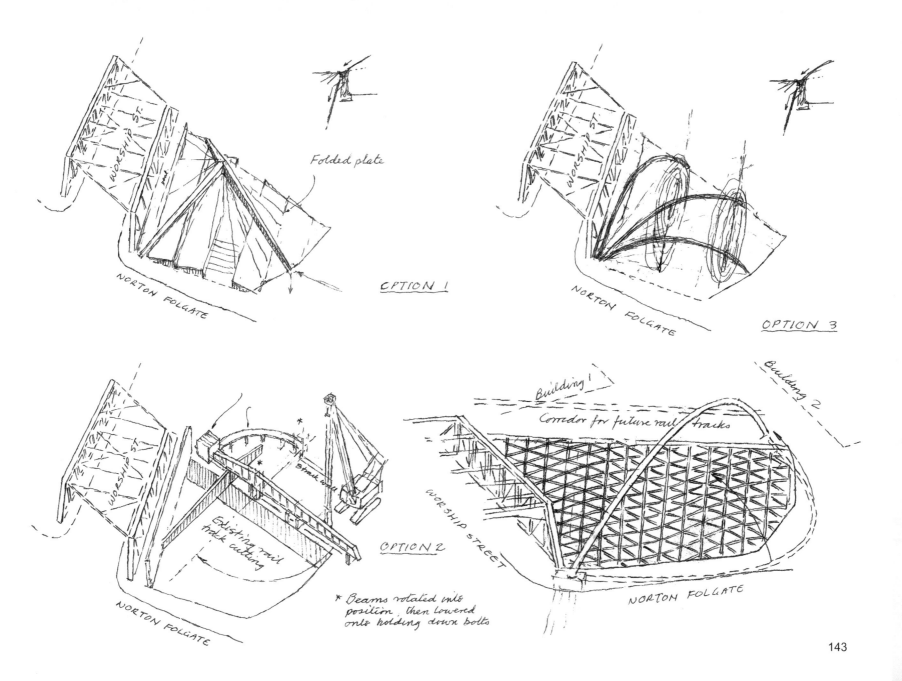

Mass transit system, Hong Kong. Mass transit systems are on a vast scale and the overall construction strategies are usually visualised at the very earliest stages using computer generated animation. These sketches were produced 25 years ago as part of a bid process but look a bit old fashioned and dated now.

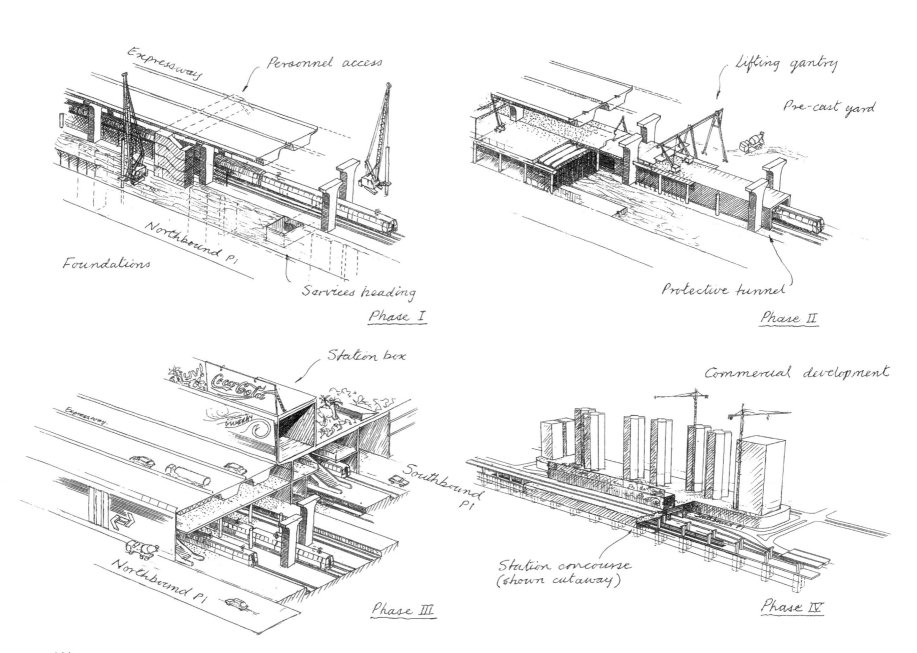

SKETCHBOOK

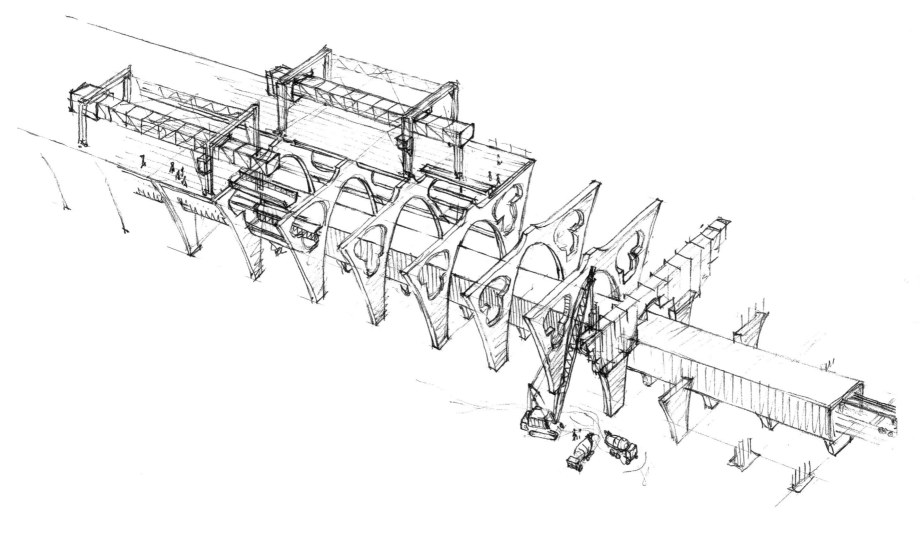

FIGURE 3.6/4

Mass transit station, Thailand. Linear build processes are not too difficult to draw, especially where primary and secondary elements are repetitive. The plan for this station was to build a box for the rail lines (and station) elevated over local traffic, construct the primary transverse elements using in-situ concrete and then to use a travelling gantry system to launch pre-cast elements at high level for the expressway. Although not built in this form, the techniques shown here, compressed in terms of timing, are not uncommon in the design and construction of mass transport systems.

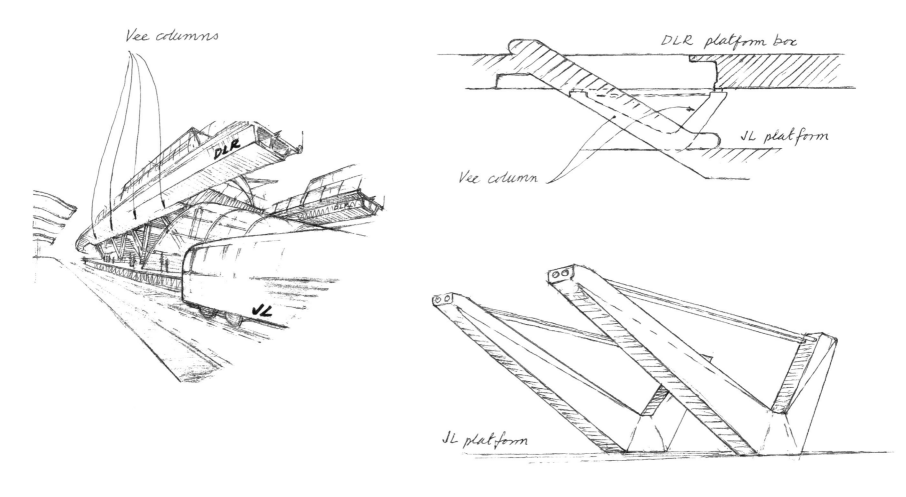

Vee columns

DLR

JL

DLR platform box

JL platform

Vee column

JL platform

FIGURE 3.6/5
Canning Town Station, Jubilee Line Extension, Canning Town, London. A key feature of the design for Canning Town Station, reconstructed as part of the Jubilee Line Extension project (q.v.), was the use of vee columns to support the Docklands Light Railway (DLR) platform boxes from the Jubilee Line (JL) platform level. The vee columns were designed to replicate the slope of the escalators with the extremities of the vees tied together to control spreading. Here the vee structures are shown disembodied, in simple cross-section and in perspective.

FIGURE 3.6/6

Canning Town Station, Jubilee Line Extension, Canning Town, London. Canning Town Station was redeveloped as part of the Jubilee Line Extension (JLE) project. It is now an interchange station between the Jubilee Line, Docklands Light Railway and the overground railway. Apart from managing existing rail services from temporary to permanent alignments, the project was greatly complicated by the presence of overhead power lines making the use of significant craneage impossible. Consequently the JLE platform structure was constructed from pre-cast segments delivered to a position clear of the power lines and slid into position using jacking systems. The segments were then effectively glued and stressed together.

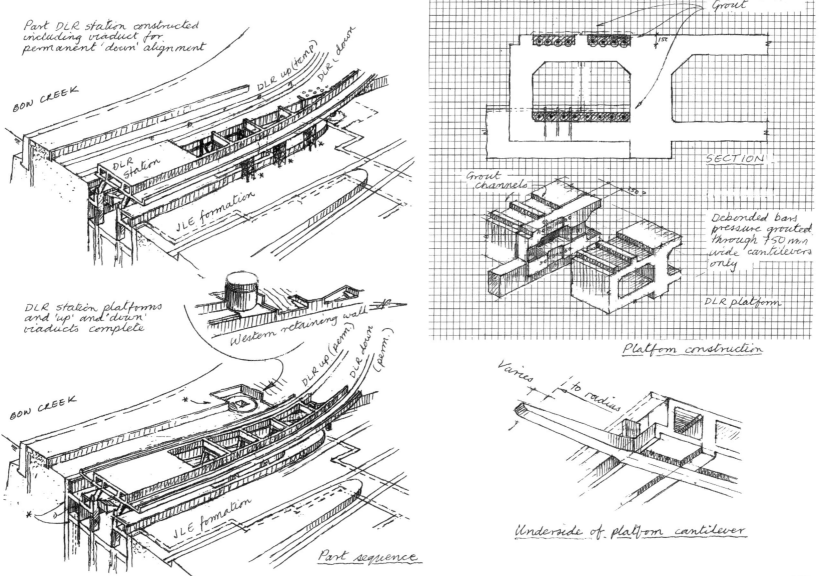

Part DLR station constructed including viaduct for permanent 'down' alignment

BOW CREEK

DLR up (temp)

DLR down

DLR Station

JLE formation

DLR station platforms and 'up' and 'down' viaducts complete

Western retaining wall

BOW CREEK

DLR up (perm.)

DLR down (perm.)

JLE formation

Part sequence

Grout

Section

Grout channels

Debonded bars pressure grouted through 750 mm wide cantilevers only

DLR platform

Platform construction

Varies to radius

Underside of platform cantilever

147

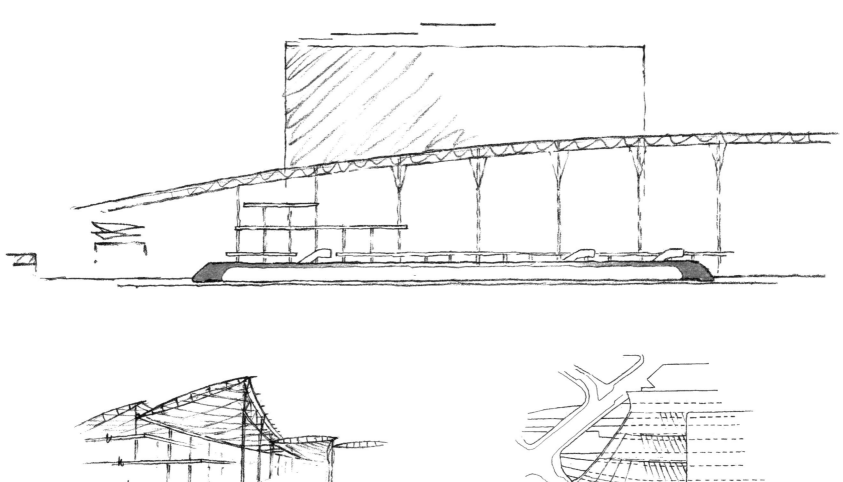

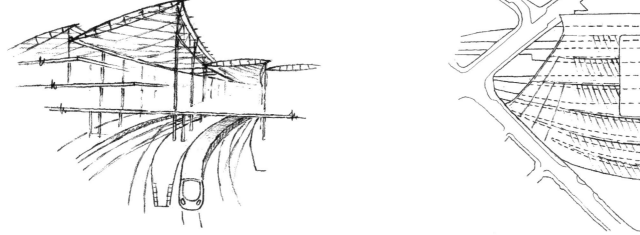

FIGURE 3.6/7
Birmingham New Street Station. Sketches for the first design stage of the ambitious scheme to redevelop Birmingham's New Street Station. Train sheds provide amazing opportunities for engineers and architects – some of our most inspiring buildings are railway stations; who cannot fail to be impressed by Paddington, Temple Meads, York, Lime Street, Waverley or the old and new King's Cross? But somehow Birmingham New Street missed out and now there was a chance to produce something cathedral-like based on the drama of arrivals and departures, on activity, the movement of the trains and the sweep of the tracks.

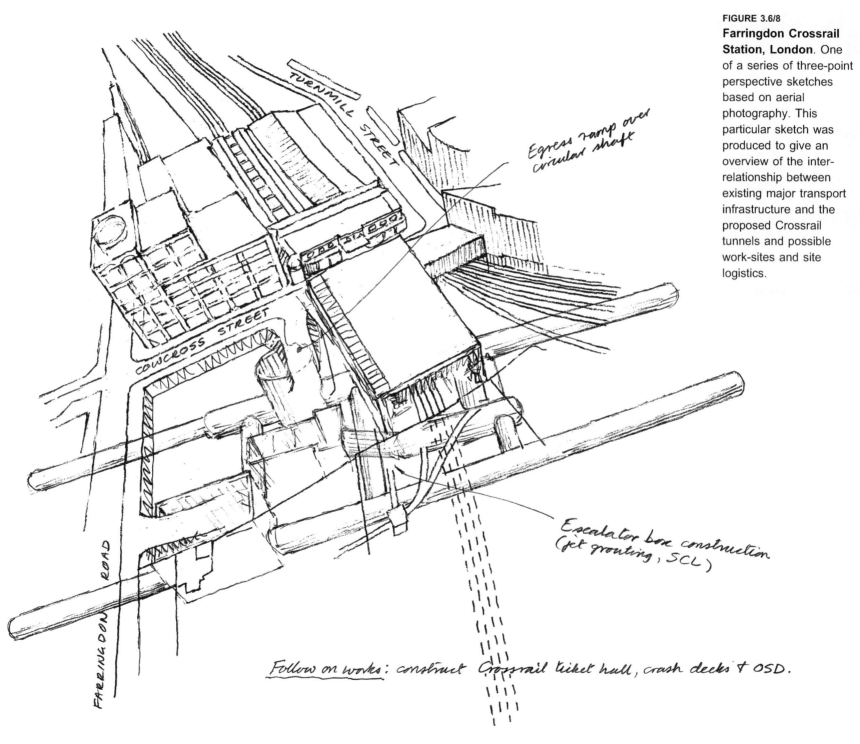

FIGURE 3.6/8
Farringdon Crossrail Station, London. One of a series of three-point perspective sketches based on aerial photography. This particular sketch was produced to give an overview of the inter-relationship between existing major transport infrastructure and the proposed Crossrail tunnels and possible work-sites and site logistics.

TURNMILL STREET

Egress ramp over circular shaft

COWCROSS STREET

FARRINGDON ROAD

Escalator box construction (jet grouting, SCL)

Follow on works: construct Crossrail ticket hall, crash decks & OSD.

Crossrail over site building, Bond Street Station, Hanover Square, London. The very first sketches are shown on the left hand side where the angle of view, what to show and what not to show are the most important considerations. Representation of massing over and beside the station box, together with basic information about how the box was to be constructed, formed the next steps in gaining a macro-scale understanding of the whole project.

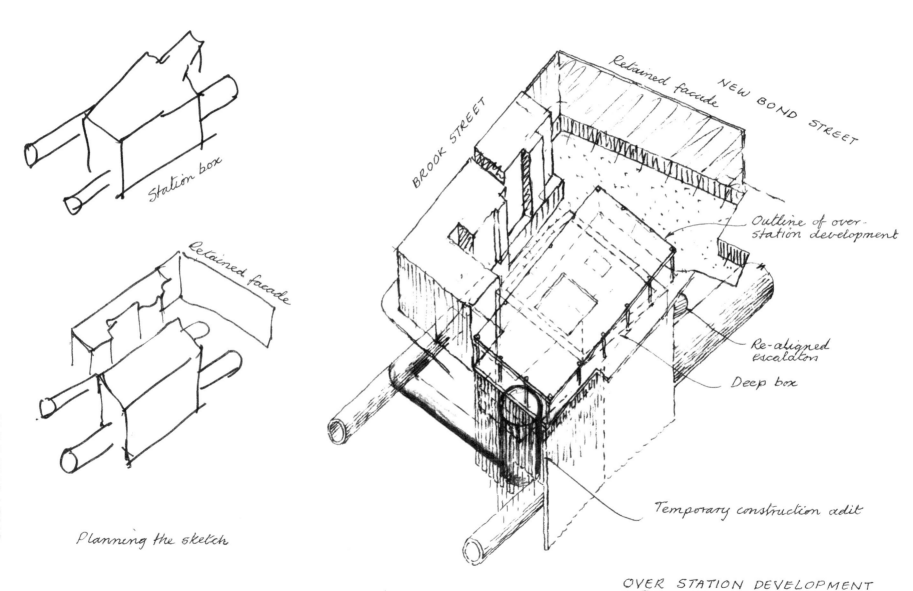

Station box

Retained facade

Planning the sketch

BROOK STREET

Retained facade

NEW BOND STREET

Outline of over-station development

Re-aligned escalaton

Deep box

Temporary construction adit

OVER STATION DEVELOPMENT

SKETCHBOOK

Station box, Crossrail, Paddington, London. These are sketches produced after thinking through the basic construction sequence. Best to get to this stage and rough something out before investing lots of time drawing something more precise.

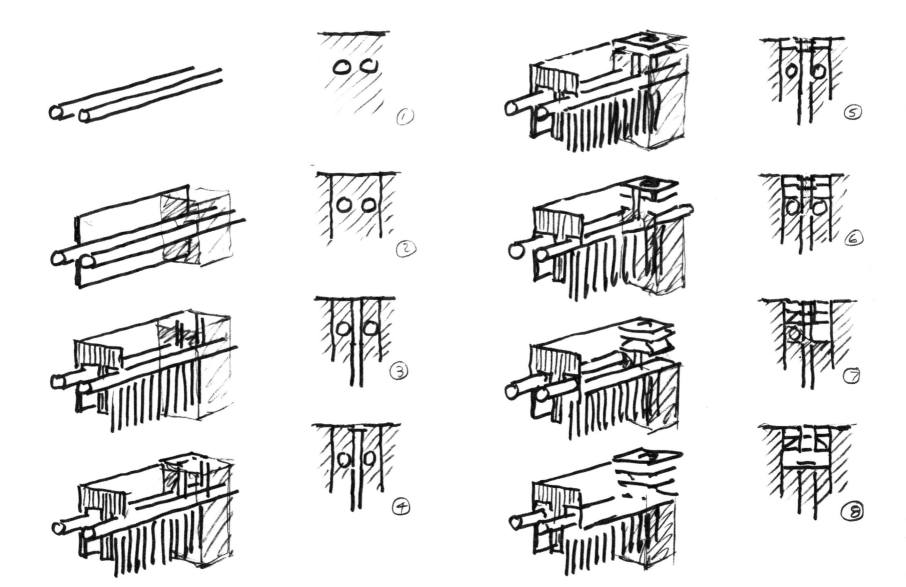

FIGURE 3.6/11

Station box, Crossrail, Paddington, London. Great opportunities presented themselves at Crossrail's Paddington Station by following to some extent the brilliant scheme developed at London Underground's Westminster station where props between the diaphragm walls are an integral part of the engineering and architectural solution. This approach is fully compatible with top-down construction, and if implemented could have been combined with a dramatic longitudinal roof light between lanes of traffic, allowing natural light to filter down to the platforms.

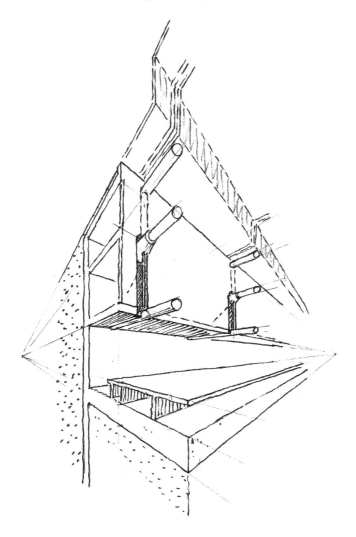

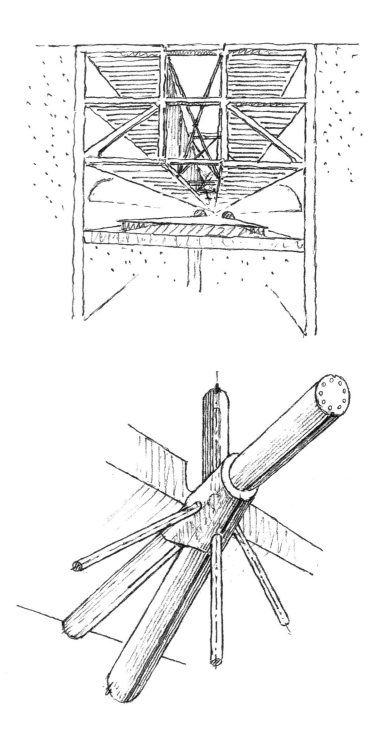

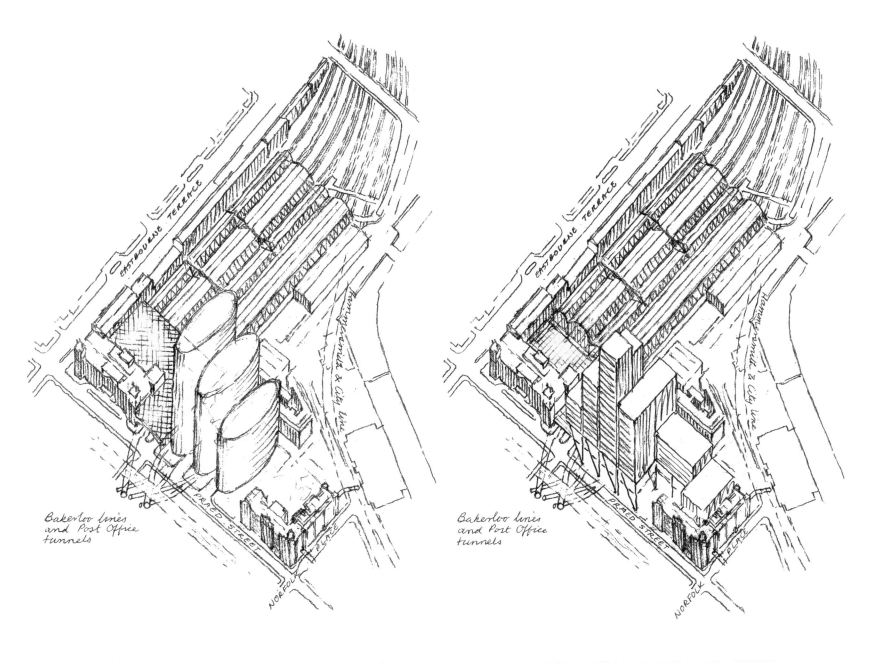

FIGURE 3.6/12

Paddington Station redevelopment. Massing of a proposed major development has to be seen in context to get a good impression of the impact on a particular locality. These are just two of a number of studies drawn quickly over a common background. The sketches can be annotated easily to show above-ground and below-ground infrastructure and to show the first ideas relating to pedestrian movement and the permeability of the various new-build options.

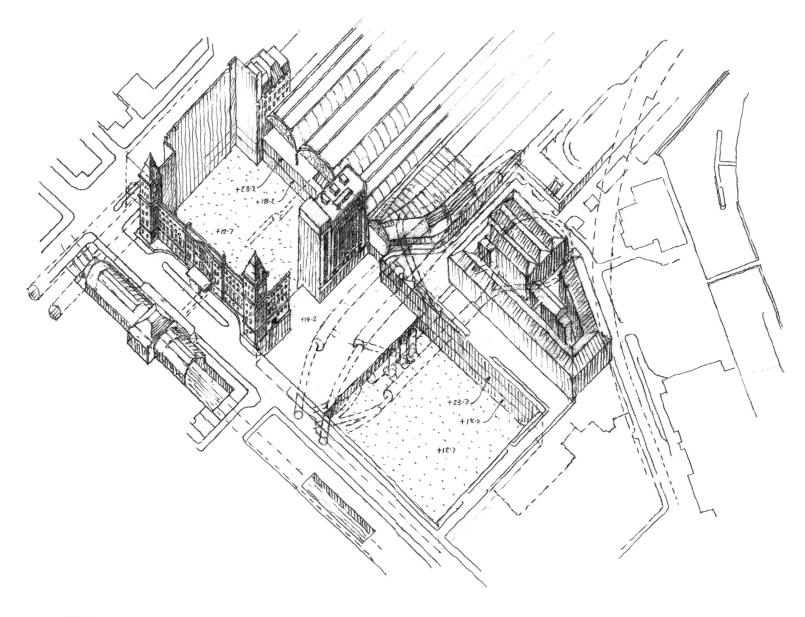

FIGURE 3.6/13

Paddington Station, London. Paddington Station is one of the most important transport hubs in London. Mainline routes, Heathrow Express, London Underground's District, Circle and Bakerloo lines are all here and Crossrail will arrive soon. Not surprisingly developers have long paid attention to the area and its increased passenger flows, especially as the Royal Mail building adjacent to the main station was available for redevelopment. This scheme shown above is one of the more ambitious and shows the mainline station being completely remodelled behind retained façades – it also shows the disused Mail Rail underground postal railway tunnels which need to be taken into account in any redevelopment.

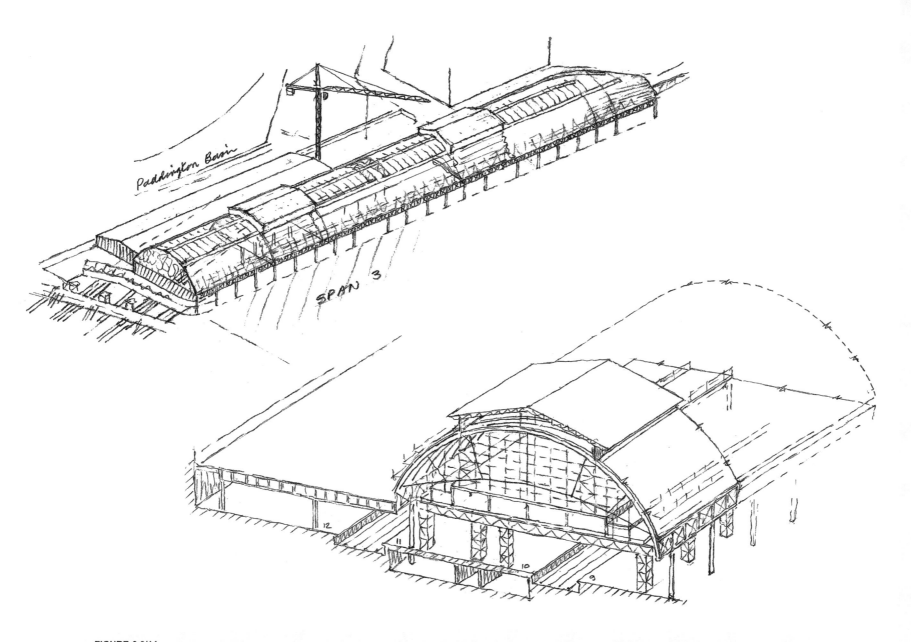

Paddington Basin

SPAN 3

FIGURE 3.6/14

Span 4 refurbishment, Paddington Station, London. The core of the station is I. K. Brunel's magnificent three-span arch roof on slender iron columns. However, in the early part of the twentieth century the station was extended, most notably by the addition of a grand fourth span to the roof structure. In recent times, span 4 had been neglected and was in much need of refurbishment to ensure it no longer appeared as a poor relation to Brunel's spans. These sketches of span 4 show work sites and working platforms positioned above the busy platforms and concourse.

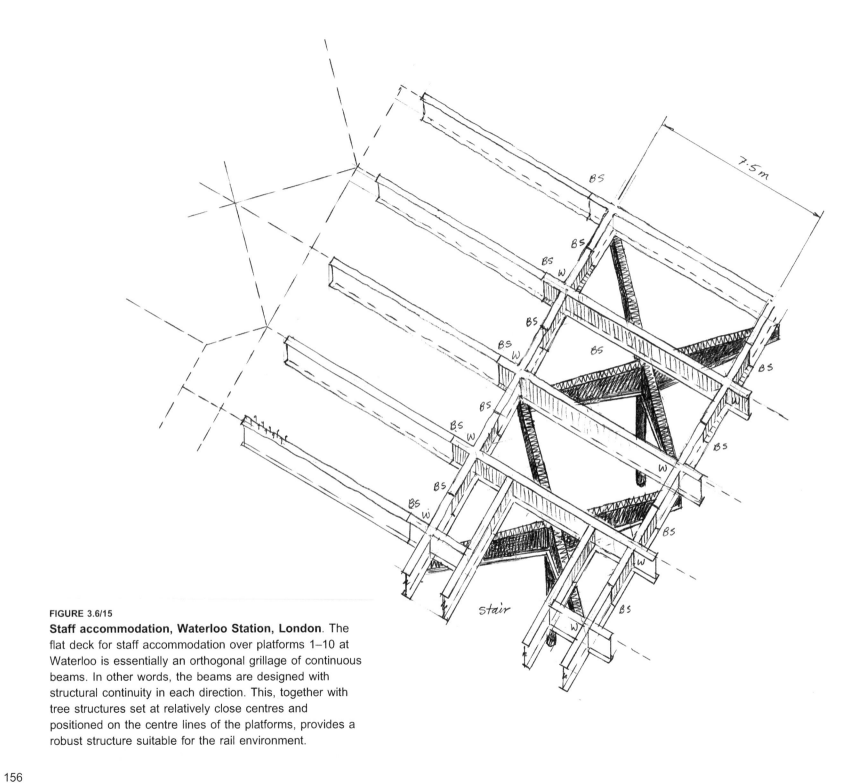

7.5m

BS

stair

FIGURE 3.6/15

Staff accommodation, Waterloo Station, London. The
flat deck for staff accommodation over platforms 1–10 at
Waterloo is essentially an orthogonal grillage of continuous
beams. In other words, the beams are designed with
structural continuity in each direction. This, together with
tree structures set at relatively close centres and
positioned on the centre lines of the platforms, provides a
robust structure suitable for the rail environment.

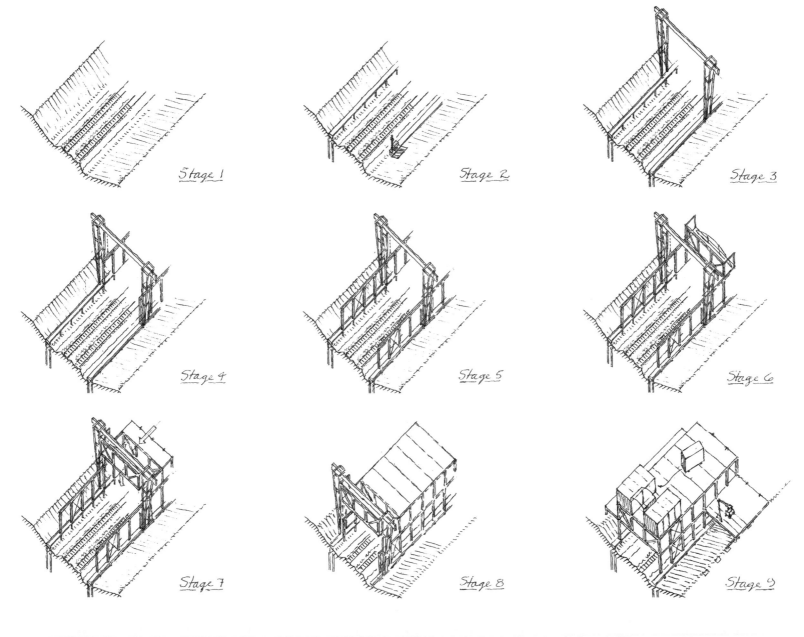

Stage 1

Stage 2

Stage 3

Stage 4

Stage 5

Stage 6

Stage 7

Stage 8

Stage 9

FIGURE 3.6/16

Rail build-over scheme. Air rights over operating railways are extremely valuable, if only we could figure out a way of constructing useable space without stopping the trains for days on end. These sequence sketches formed part of an in-depth study showing how prefabricated elements could be slid into position from a single launch site. The slide time in each cycle is short and could take place during 'engineering hours' which are available on most routes.

3.7   PORTS AND MARINE

It is always a welcome change to be able to draw or sketch
something to do with ships and the sea, but the opportunities
often belong to civil rather than structural engineers or architects.
There is a new language to learn: jetties, wharves, breakwaters
and cut-waters, dolphins and canting brows, and of course new
machinery and techniques to understand and illustrate. To
develop any useful ideas it is good to know what can be achieved
using dredging techniques, or how cofferdams are constructed
and how marine piles can be driven from barges or spud barges
– a great bit of specialist terminology: a jack-up barge, a spud
barge has four legs (called spud poles) that telescope to the river
bed and provide a wharf and usually have a deployable ramp with
side rails.

Structural engineers and architects are most likely to come into
contact with our marine and civil cousins where land meets water.
We are then all dealing with moorings and fenders, with berthing
loads and ship impact loads and with closely allied structures
such as dockside logistics facilities, cranes and floating amenities.

When it comes to offshore structures, dock, harbour and sea
walls, then these are often of a different scale and the realm of
the specialist engineers and naval architects.

SKETCHBOOK

FIGURE 3.7/1

Ferry terminus, Hamburg. This sketch shows a single bay of a ferry terminus once proposed for the Port of Hamburg. It was an ambitious scheme with Wil Alsop – this version was never developed beyond the early concept stage, and was based on a central structure consisting of two jump-formed shells connected by pre-cast concrete vierendeel beams. This configuration gave almost unlimited opportunities for link bridges, escalators and walkways within the longitudinal core itself while allowing a degree of spatial transparency. Between the core and the outer envelope, floor beams were designed with top and bottom slabs taking balanced push pull loads through the vierendeel beams. Structure within the outer envelope would have been used to provide deflection control and effectively turn the cantilevers into much more efficient propped cantilevers.

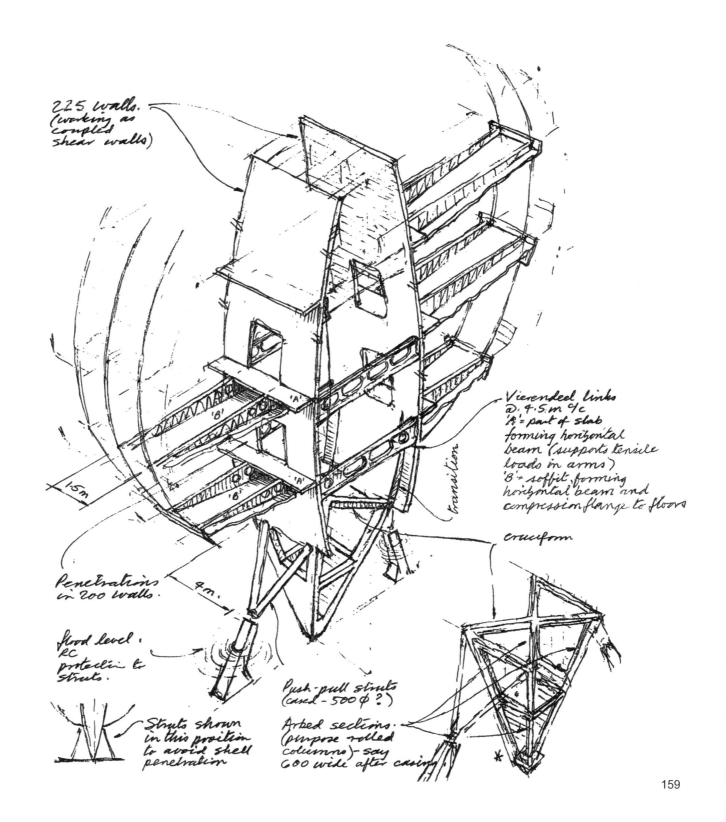

225 walls.
(working as
coupled
shear walls)

Vierendeel links
@ 4.5 m %c
'A' = part of slab
forming horizontal
beam (supports tensile
loads in arms)
'B' = soffit forming
horizontal beam and
compression flange to floors

transition

cruciform

15 m

Penetrations
in 200 walls.

4 m.

flood level.
RC
protection to
struts.

Push-pull struts
(cased - 500 φ ?)

Arbed sections.
(purpose rolled
columns) - say
600 wide after casing.

Struts shown
in this position
to avoid shell
penetration

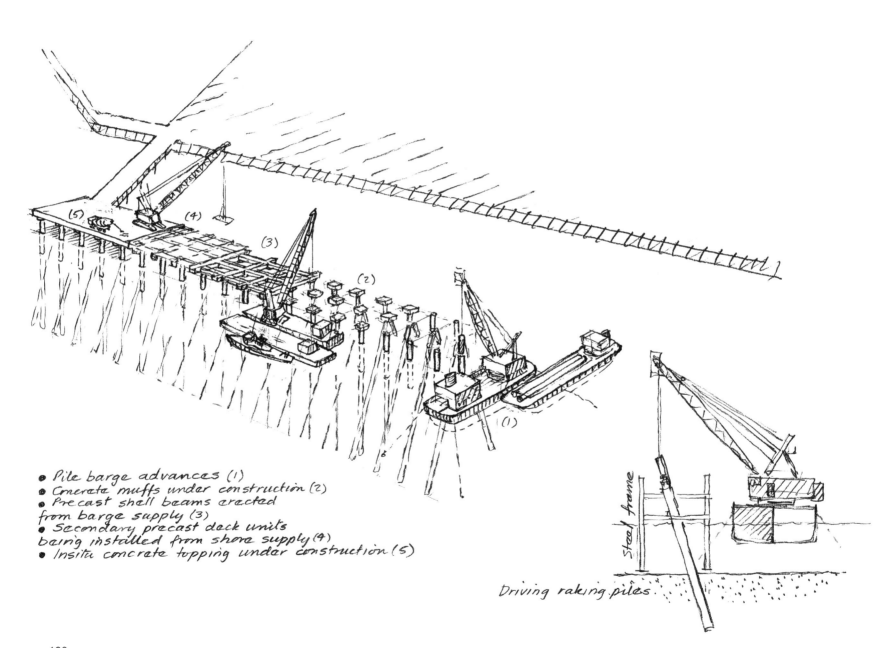

FIGURE 3.7/2

Power station, the Philippines. In the early 1990s, the economy of the Philippines was apparently being held back by an ageing national grid and as a consequence the World Bank suggested that the government attract monetary interest by approving a build-own-transfer scheme for power project developments. The result was this power station, now one of the largest coal power plants in the Philippines. The sketch is part of a marine works construction sequence.

- Pile barge advances (1)
- Concrete muffs under construction (2)
- Precast shell beams erected from barge supply (3)
- Secondary precast deck units being installed from shore supply (4)
- Insitu concrete topping under construction (5)

Steel frame

Driving raking piles.

SKETCHBOOK

160

3.8   STADIA

In some ways, stadia are like airport terminus buildings: they are meant to be impressive and awe-inspiring. Once just a collection of cantilevered canopies or glorified sheds, they are now so often an integral part of the brand and prestige of a sporting venue. Sports stadia of unprecedented complexity and scale are being built, incorporating landmark architecture and innovative structural engineering.

Effective design of large span structures with complex external envelopes is often dictating lightweight structural forms, making these roofs highly sensitive to loading from wind and snow. In order for sophisticated 3D structural modelling to deliver efficient structural design, accurate loading scenarios are required for wind that account for complex fluid–structure interactions, including wind driven dynamic effects. Wind related environmental impact also affects pitch microclimate, spectator comfort and external microclimate and all require careful consideration from an early stage in design.

But the real fun comes early on, when everyone is searching for the right solution or for something unique, a new way of producing a dramatic design or just something that is different. It is easy to sketch ideas but hard to take all the inevitable constraints into account. Fulham Football Club's ground at Stevenage Road beside the Thames in London is typical of an ongoing design challenge. Not only is the stadium beside the river, but it is surrounded by residential properties and narrow streets, and is without doubt in a highly sensitive area from a planning point of view. At least, its context might eventually inform the design, by respecting its neighbours, taking account of existing stands and their heritage value, and perhaps, in terms of constructability, taking advantage of the much under-used River Thames.

SKETCHBOOK

161

FIGURE 3.8/1

Lansdowne Road, Dublin. Searching for something new in the design of stadia is not easy. Two of these sketch proposals for the Lansdowne Road competition were based on the conventional hockey stick approach and one on triangulated arches positioned above the canopy and spanning the length of the pitch. The fourth design, bottom right, was a little more inventive, and borrowed technology from cable stayed bridge designs.

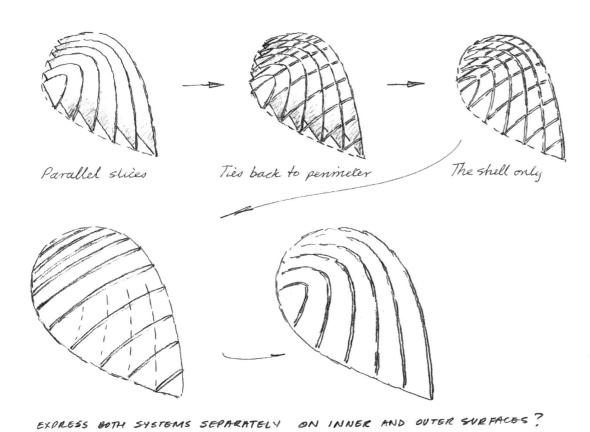

Parallel slices   Ties back to perimeter   The shell only

EXPRESS BOTH SYSTEMS SEPARATELY ON INNER AND OUTER SURFACES ?

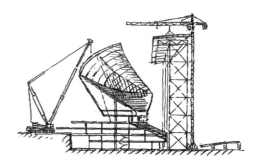

ERECTION Phase 1

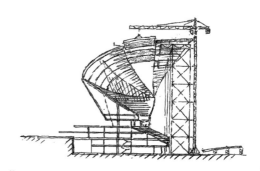

ERECTION Phase 2

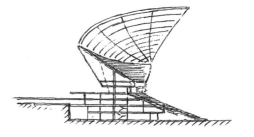

ERECTION Phase 3

FIGURE 3.8/2

Stadium competition, Casablanca, Morocco. This competition design was based on shell-like canopies, one on each side of a rectangular sports arena, and constructed as free-standing structures. A way of structuring each canopy was developed by studying the surface of any one of the canopies, by cutting slices in various directions, and deciding whether the interconnection of the cut lines could suggest a viable, constructible system of load sharing frames.

SKETCHBOOK

163

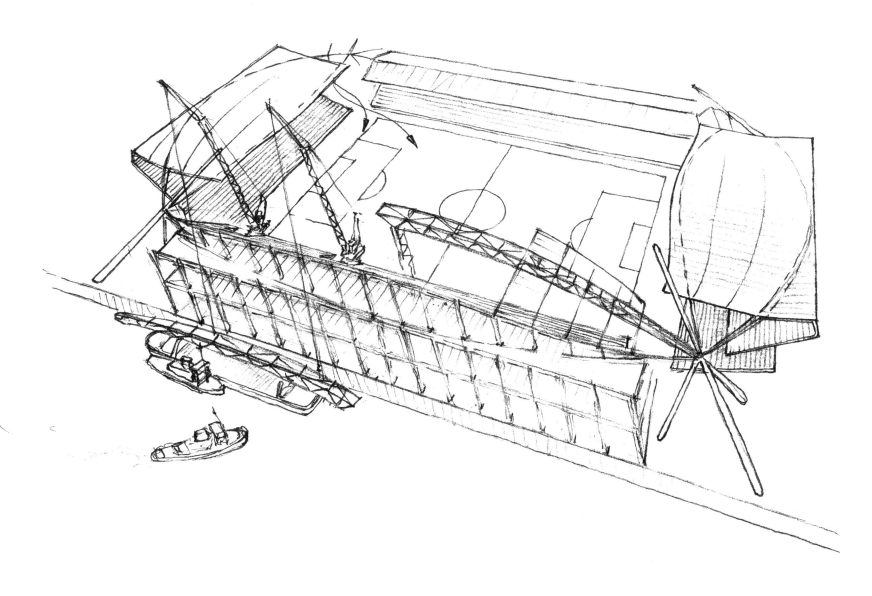

FIGURE 3.8/3

Fulham Football Club, Craven Cottage, London. The land-side area around Fulham Football Club's ground at Craven Cottage is residential and the streets are narrow. The River Thames is therefore the best way of delivering large prefabricated elements to site for the redevelopment of the Riverside Stand. Sketch reproduced by kind permission of Fulham Football Club.

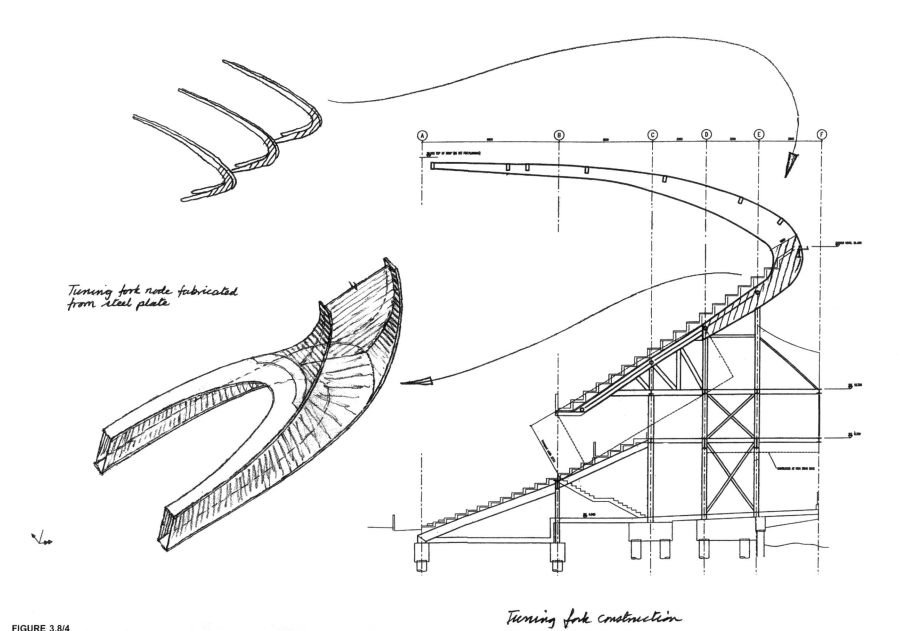

Tuning fork node fabricated from steel plate

Tuning fork construction

FIGURE 3.8/4

Fulham Football Club, Craven Cottage, London. Another early scheme, this time with bifurcated hockey stick cantilevers. The shape of the node, generated by the position of the vomitory (access to the terrace seating), is the key component. It is hand drawn and superimposed on a two-dimensional CAD drawing to illustrate its fabrication from single curvature welded steel plates with size and weight designed to suit road transport. Sketch reproduced by kind permission of Fulham Football Club.

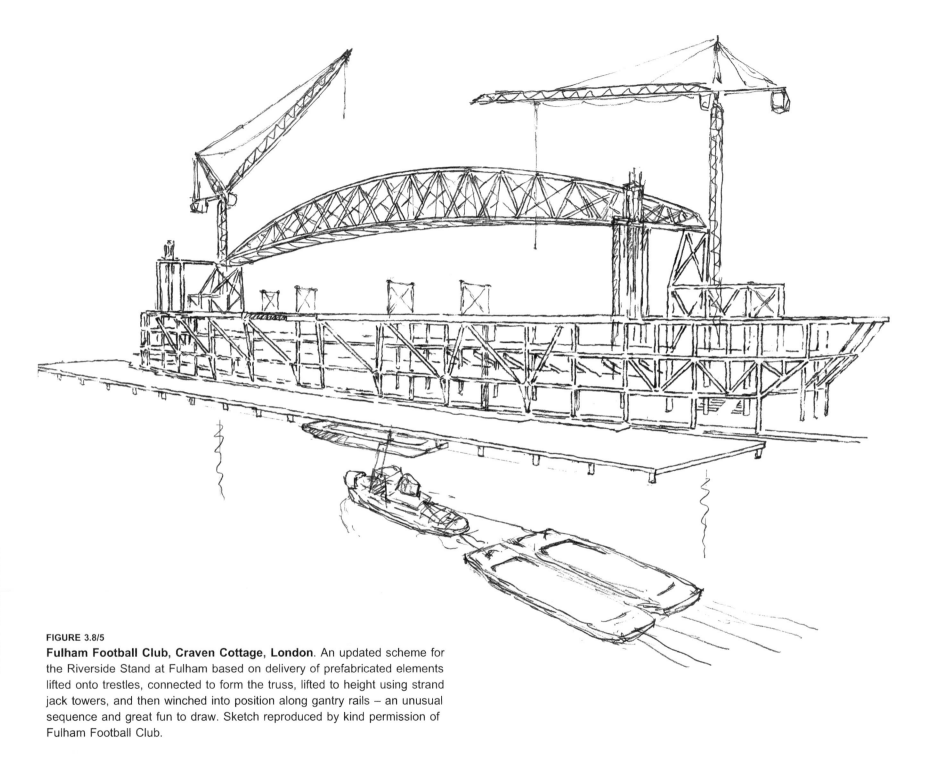

FIGURE 3.8/5
Fulham Football Club, Craven Cottage, London. An updated scheme for
the Riverside Stand at Fulham based on delivery of prefabricated elements
lifted onto trestles, connected to form the truss, lifted to height using strand
jack towers, and then winched into position along gantry rails – an unusual
sequence and great fun to draw. Sketch reproduced by kind permission of
Fulham Football Club.

3.9 EXISTING BUILDINGS

Existing buildings, and especially historic buildings, are great to sketch, just because at the end of it, and with a fair wind, you have an image that sums up your interpretation of what you see. If nothing more, producing the sketch is satisfying in its own right and you get the same sort of pleasure that an artist gets from drawing or painting. There's also the opportunity to add a bit of character to the building rather than drawing it with the cold precision we often have to use.

But in terms of engineering and architecture, we mustn't lose sight of why we draw and sketch in the first place. Often it's to gain an understanding of a particular building or structure. Here it is good to mentally edit the information available. Show broad outline – not too much detail, trace load paths, explain movement that may have occurred (arrows are useful – what would we have done if arrows had not been invented?). There's the opportunity to show context, adjoining buildings and roads, basements and tunnels if they exist, and of course geology.

Gathering information can be difficult. Walking the site and photography helps and so do web-based aerial and street views, but you can't beat construction drawings or as-built drawings. There's an important difference: construction drawings are more commonly available but may not represent the as-built building so must always be used with a degree of caution.

It is well worth searching for archival material – search by the engineer's or architect's name, try biographical dictionaries, online databases, libraries and specialist societies.

Even if you have found good base material, you still have to decide what to show and what not to show: is your objective to describe broad outline, how the building stands up, which walls contribute to stability, or is it a particular part or special details that you are interested in? Then you have to plan how to illustrate your findings; cutaway drawings are very persuasive and sections are invaluable – engineers and architects should always draw sections.

FIGURE 3.9/1

Royal College of Music, London. Alterations and additions to existing buildings present their own special challenges with respect to gaining rapid understanding of scale and complexity of a proposed project. This is especially true of schemes involving older buildings where form and layout do not follow a simple pattern. Drawings available at bid stage are sometimes inadequate and the best approach is often to make use of aerial photographs and then use colour or shading to pick out the areas to be refurbished or infilled.

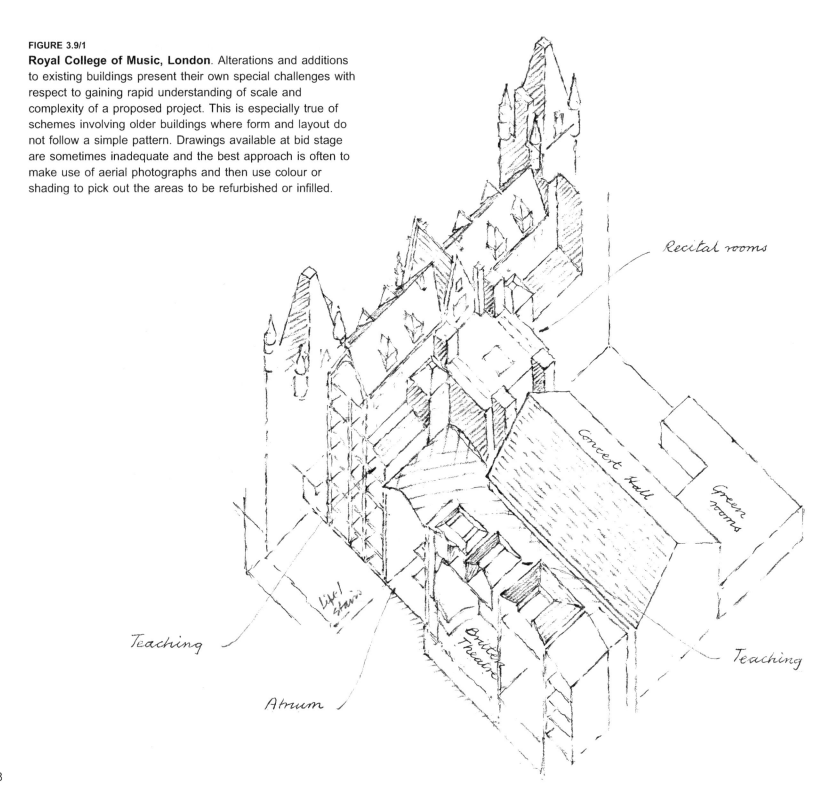

Recital rooms

Concert Hall

Green rooms

Lift/Stair

Teaching

Britten Theatre

Teaching

Atrium

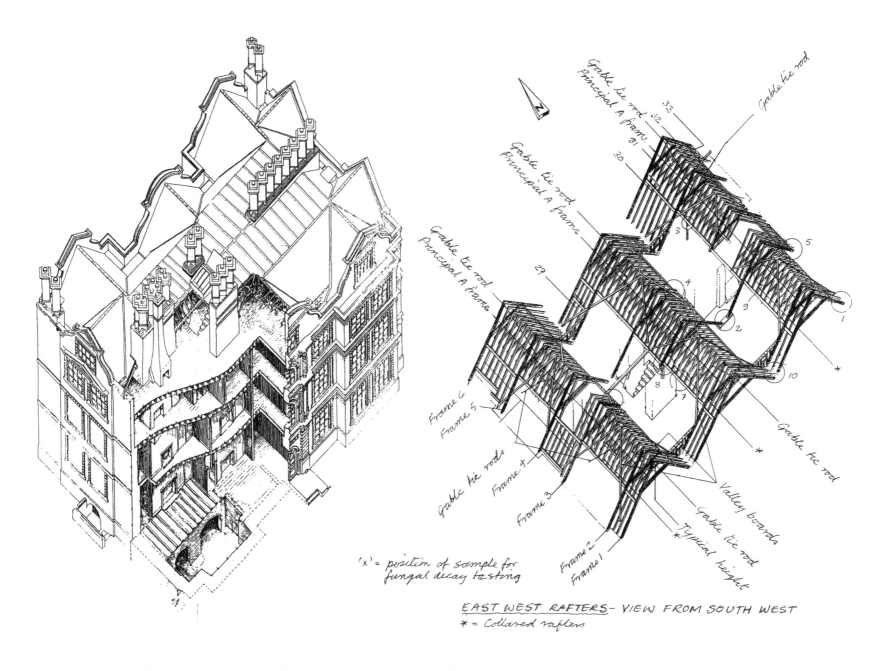

Gable tie rod
Principal A frame 33 32 31

Gable tie rod
Principal A frame 30

Gable tie rod
Principal A frame

Gable tie rod
Principal A frame 27

Gable tie rod

Gable tie rod

N

3 5 4 9 2 1 10 *

8 7

Frame 6
Frame 5

Gable tie rods
Frame 4
Frame 3

Frame 2
Frame 1

Valley boards
Gable tie rod
Typical height

'x' = position of sample for
fungal decay testing

EAST WEST RAFTERS – VIEW FROM SOUTH WEST
* = Collared rafters

FIGURE 3.9/2

Kew Palace, Kew Gardens, London. Cutaway drawings help to get a feel of how a complex building has been put together – drawing and understanding go together; you have to find answers to questions before you can draw it. The roof at Kew Palace is particularly complex and of course no drawings existed showing how it worked structurally. Eventually, east–west timbers and north–south timbers were shown on separate drawings for the sake of clarity. (East–west only shown here.)

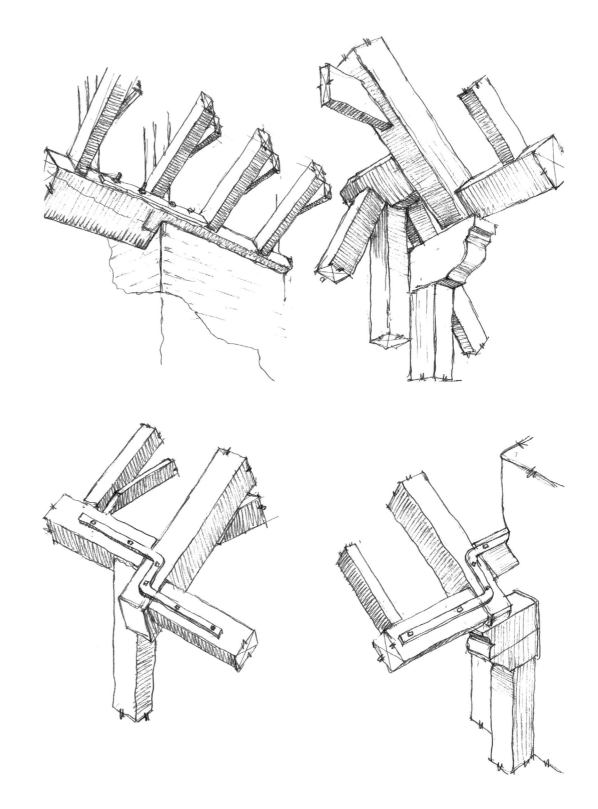

FIGURE 3.9/3
Kew Palace, Kew Gardens, London.
Photography in a dark roof space leaves a lot to be desired. So often missed detail and shadows mean a return visit is inevitable. If time permits it's so much better to at least supplement photographs with on-site sketches. However, the over-riding memory of drawing these timber junction details at Kew in mid-winter was the intense cold and numb fingers.

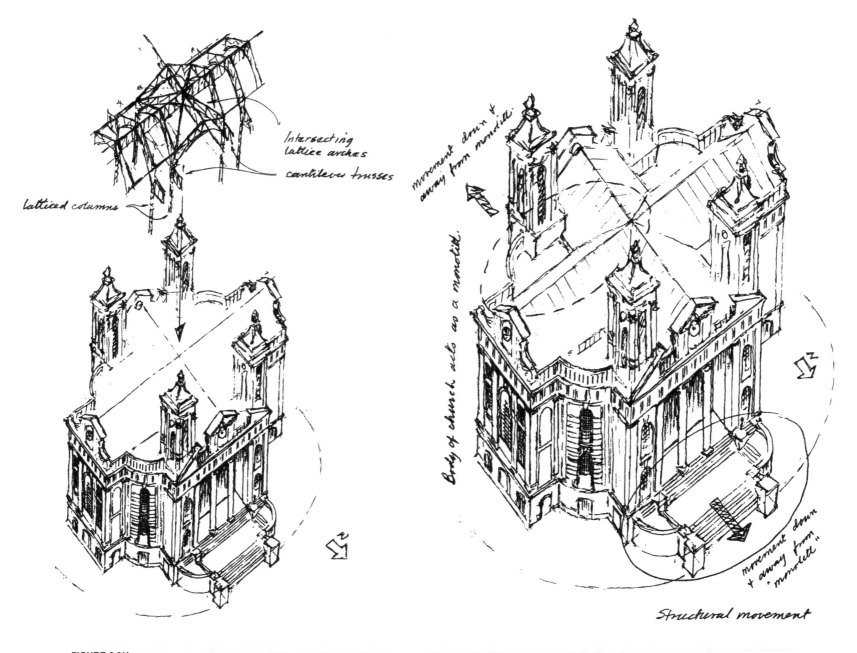

Intersecting
lattice arches

cantilever trusses

latticed columns

movement down &
away from monolith

Body of church acts as a monolith

movement down & away from "monolith"

Structural movement

FIGURE 3.9/4

St John's Concert Hall, Smith Square, London. St John's Concert Hall is an interesting building restored and re-roofed after the Second World War but showing signs of distress due to ground movement. There was no need for this drawing to be precise in every detail; it was produced to illustrate historic interventions (the 'new' roof) and macro-scale movements.

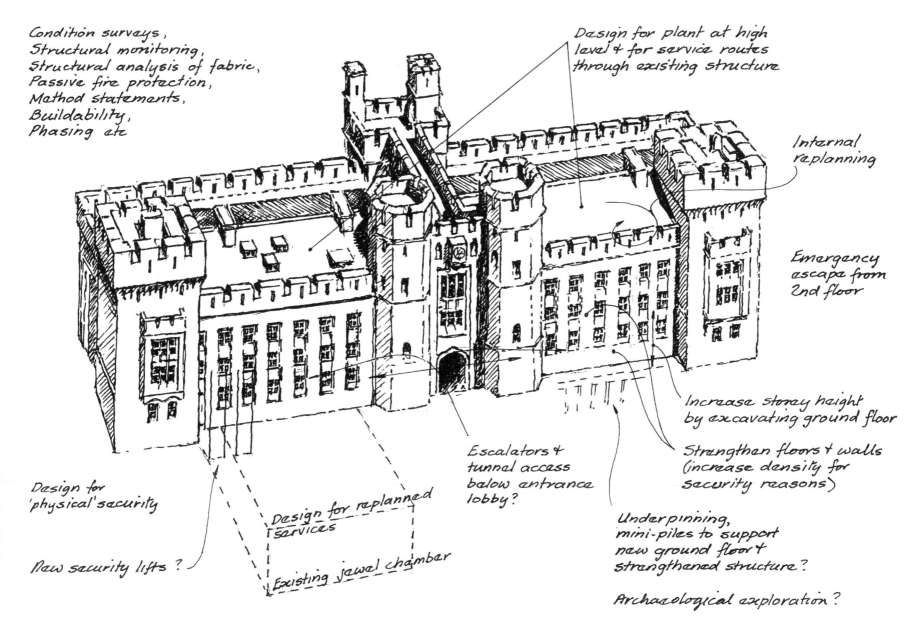

FIGURE 3.9/5

The Jewel House, the Tower of London, London. Some older buildings are just nice to draw and somehow deserve a less precise line style. This sketch of the Jewel House at the Tower of London, drawn from simple plans and elevations, was produced as part of a bidding process to illustrate the team's understanding of the scope of works to be undertaken during a major renovation project.

Condition surveys,
Structural monitoring,
Structural analysis of fabric,
Passive fire protection,
Method statements,
Buildability,
Phasing etc

Design for plant at high
level & for service routes
through existing structure

Internal
replanning

Emergency
escape from
2nd floor

Increase storey height
by excavating ground floor

Strengthen floors & walls
(increase density for
security reasons)

Escalators &
tunnel access
below entrance
lobby?

Underpinning,
mini-piles to support
new ground floor &
strengthened structure?

Design for replanned
services

Archaeological exploration?

Design for
'physical' security

Existing jewel chamber

New security lifts?

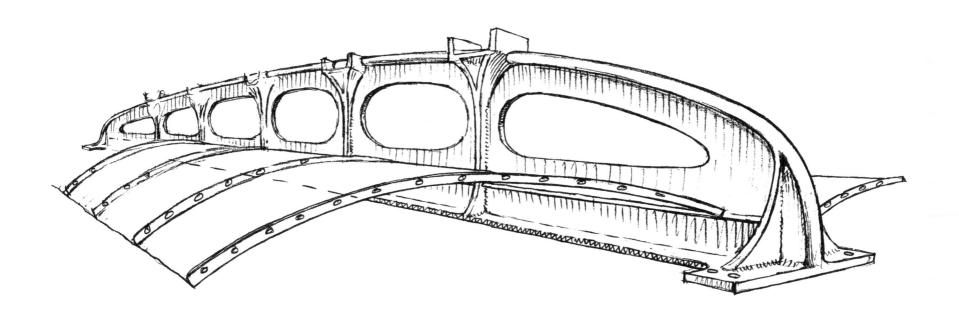

FIGURE 3.9/6
The King's Library, the British Museum, London. Victorian engineer John Rastrick was responsible for 40 and 50ft spanning cast girders used in the main floor structure of the gallery at the Museum known formerly as the King's Library. These beautiful castings, sadly completely hidden within the floor construction, can only be viewed by intrepid surveyors (and engineers) crawling through dusty floor voids. They must have been a splendid sight travelling down the canals on barges from the Midlands.

The King's Library, the British Museum, London. Change of use from library to gallery space suitable for the display of heavy objects, meant that the original cast iron beams could no longer be relied upon. The sketch shows supplementary trusses positioned either side of the casting. The arrangement was designed to be 'reversible' – in other words, the original structure was left unaltered so that at some future date, the supplementary trusses could be removed or replaced if necessary without damaging the original fabric.

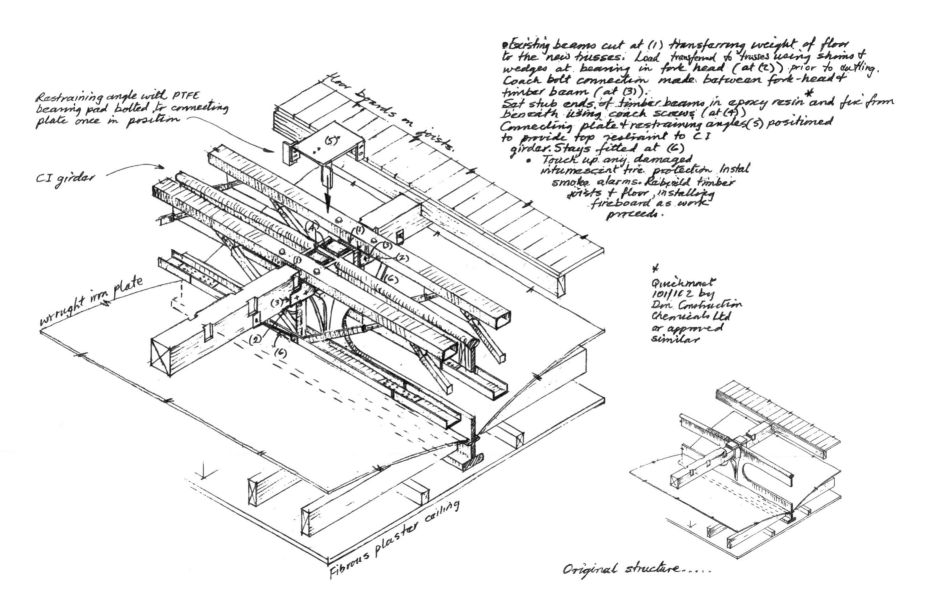

SKETCHBOOK

FIGURE 3.9/8

Kinnaird House, Pall Mall East, London.

2D sections were used to show the main features of the existing building and the rebuild proposals – sometimes the simplest is the best. Drawing sections like this is a natural way of excluding unnecessary and confusing detail.

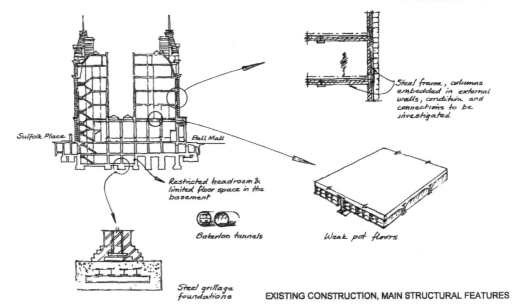

The elements of the existing building requiring special structural attention are illustrated below:

Suffolk Place

Pall Mall

Steel frame, columns embedded in external walls, condition and connections to be investigated

Restricted headroom & limited floor space in the basement

Bakerloo tunnels

Weak pot floors

Steel grillage foundations

EXISTING CONSTRUCTION, MAIN STRUCTURAL FEATURES

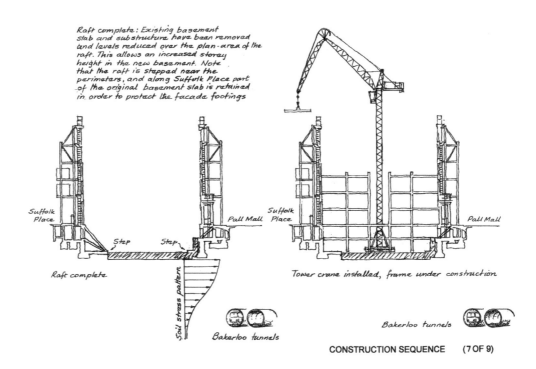

Raft complete: Existing basement slab and substructure have been removed and levels reduced over the plan-area of the raft. This allows an increased storey height in the new basement. Note that the raft is stepped near the perimeters, and along Suffolk Place part of the original basement slab is retained in order to protect the facade footings

Suffolk Place

Pall Mall

Step

Step

Raft complete

Soil stress pattern

Bakerloo tunnels

Suffolk Place

Pall Mall

Tower crane installed, frame under construction

Bakerloo tunnels

CONSTRUCTION SEQUENCE   (7 OF 9)

FIGURE 3.9/9

Kinnaird House, Pall Mall East, London. Almost the perfect project for façade retention, an island site, four sided, almost square on plan and pavement space available for external restraint frames. The frames could be supplemented by minimal internal braces positioned below the corner turrets. The only issue was whether to retain the mansards and chimney stacks. A decision was made to dismantle and rebuild, which is often the outcome on projects of this nature and scale.

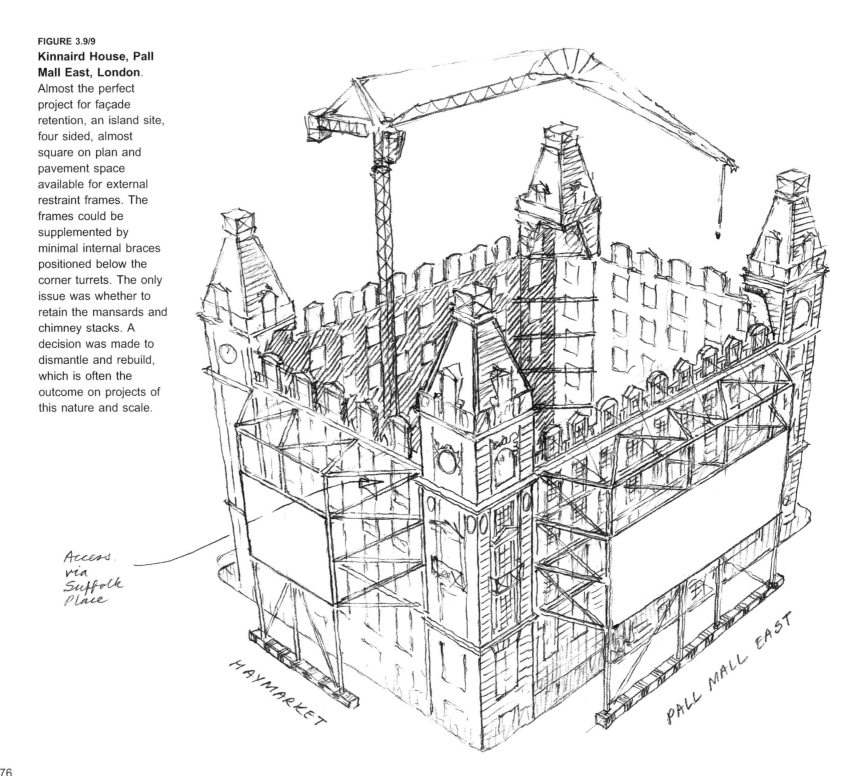

Access via Suffolk Place

HAYMARKET

PALL MALL EAST

FIGURE 3.9/10

King William Street, London. The façades at King William Street did not lend themselves to simple retention systems. Limited load bearing capacity at pavement level precluded the use of a fully cantilevered external system. Instead, internal temporary works towers were proposed which are difficult to build while the existing structure is in place. They are difficult to build around and are awkward to dismantle after the new frame is erected.

1) Posts carry self weight of restraint frame only - spreader foundations required at pavement level

2) Wind girders at alternate levels

3) Internal facade restraint towers

4) Knee braces

5) Shores to retaining walls

6) Rear wall restraint system

7) Rear wall wind girders

8) Lateral bracing.

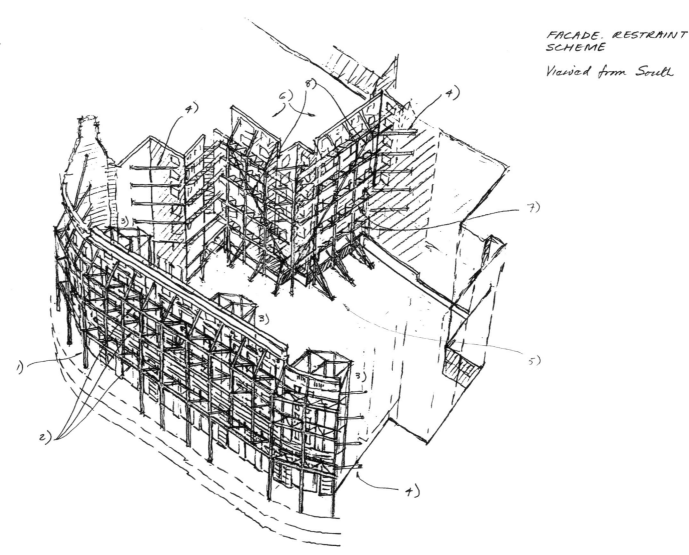

FACADE. RESTRAINT SCHEME

Viewed from South

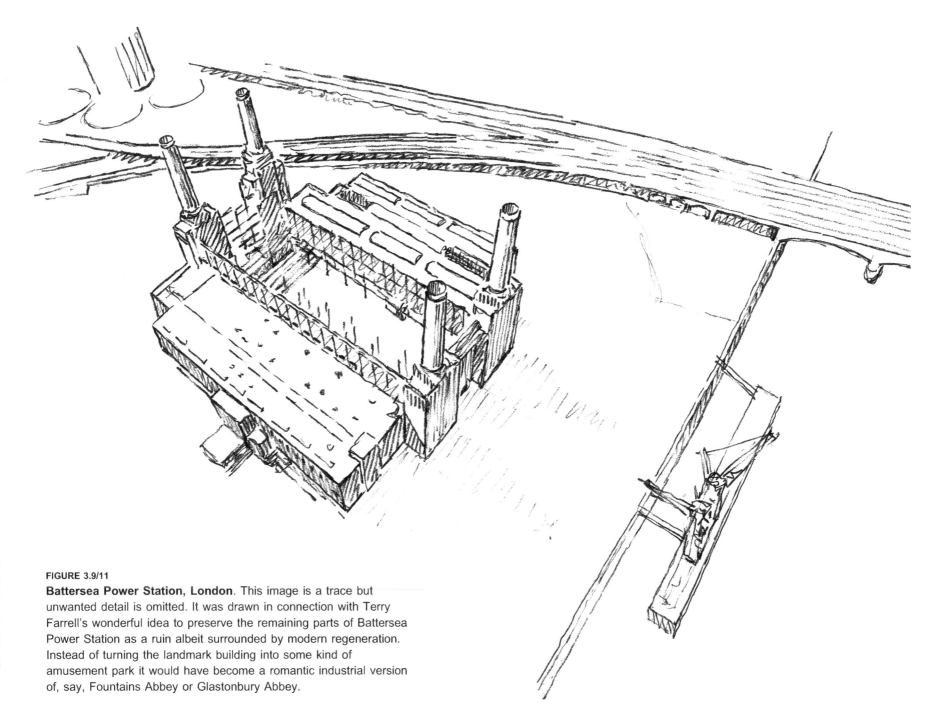

FIGURE 3.9/11

Battersea Power Station, London. This image is a trace but unwanted detail is omitted. It was drawn in connection with Terry Farrell's wonderful idea to preserve the remaining parts of Battersea Power Station as a ruin albeit surrounded by modern regeneration. Instead of turning the landmark building into some kind of amusement park it would have become a romantic industrial version of, say, Fountains Abbey or Glastonbury Abbey.

FIGURE 3.9/12

Harrods, Knightsbridge, London. Secant piled lift shafts were constructed in a light-well of the Harrods main building to take freight lifts well below ground. This was part of a scheme to link the main building to storage and loading bays in the redeveloped site at Knightsbridge Crown Court east of Basil Street via a new tunnel. Extensive piling works were needed to support alterations to the light-well and the main building.

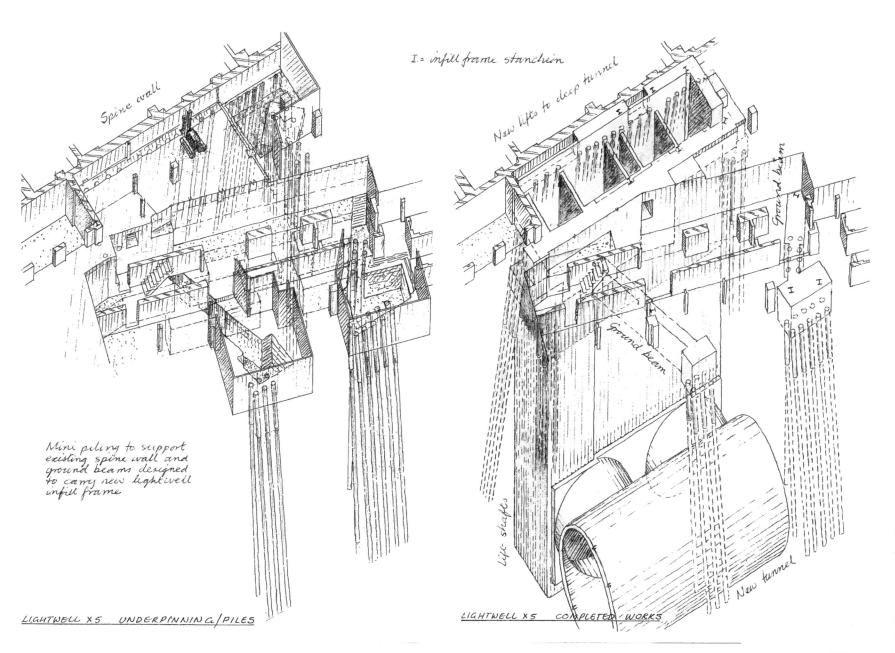

Spine wall

I = infill frame stanchion

New lifts to deep tunnel

Ground beam

Ground beam

Mini piling to support existing spine wall and ground beams designed to carry new lightwell infill frame

Lift shafts

New tunnel

LIGHTWELL X5   UNDERPINNING/PILES

LIGHTWELL X5   COMPLETED WORKS

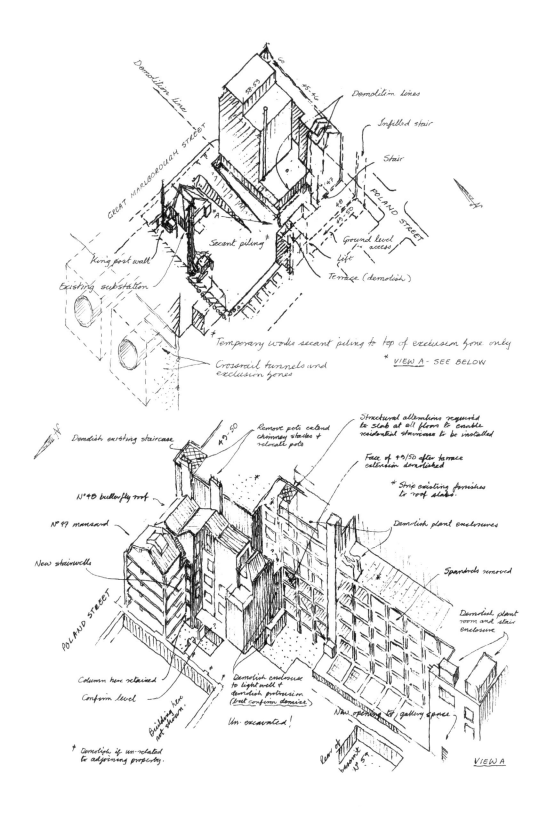

Demolition line

GREAT MARLBOROUGH STREET

58-53

60

45-46

Demolition lines

Infilled stair

Stair

POLAND STREET

Ground level access

Lift

Terrace (demolish)

Secant piling ‡

King post wall

Existing substation

‡ Temporary works secant piling to top of exclusion zone only

Crossrail tunnels and exclusion zones

* VIEW A - SEE BELOW

Demolish existing staircase

49-50

Remove pots extend chimney stacks & relocate pots

Structural alterations required to slab at all floors to enable residential staircase to be installed

Face of 49/50 after terrace extension demolished

* Strip existing finishes to roof slabs.

N° 48 butterfly roof

N° 47 mansard

New stairwells

Demolish plant enclosures

Spandrels removed

POLAND STREET

Demolish plant room and stair enclosure

Column here retained

Confirm level

Building here not shown.

* Demolish enclosure to light well + demolish protrusion (but confirm demise)

Un-excavated!

New opening to gallery space

rear of basement N° 51

* Demolish if un-related to adjoining property.

VIEW A

55–57 Great Marlborough Street, London. 55–57 Great Marlborough Street is another of those urban sites that suffer from all kinds of constraints – confined space, interface with adjoining buildings, party wall issues, new-build height constraints, possible archaeological remains, Crossrail running directly under the site, an existing sub-station, next to a leading sound recording studio – the list seems endless. Drawing the site after demolition is a good start (after deciding which view to use and which buildings not to draw 'for clarity'). Then thinking through a notional build sequence gives everyone involved a better idea of the challenges ahead.

SKETCHBOOK

FIGURE 3.9/14

Newfoundland, Canary Wharf, London. An unusual constraint in London Docklands is the existence of the old Victorian dock walls known affectionately as the banana walls. The banana walls are on the statutory list of buildings of special architectural or historic interest and are classified as Grade 1 and cannot be demolished, extended, or altered without special permission from the planning authorities. Here a primary load is diverted to a group of piles clear of the wall and beyond the London Underground's Jubilee Line exclusion zone.

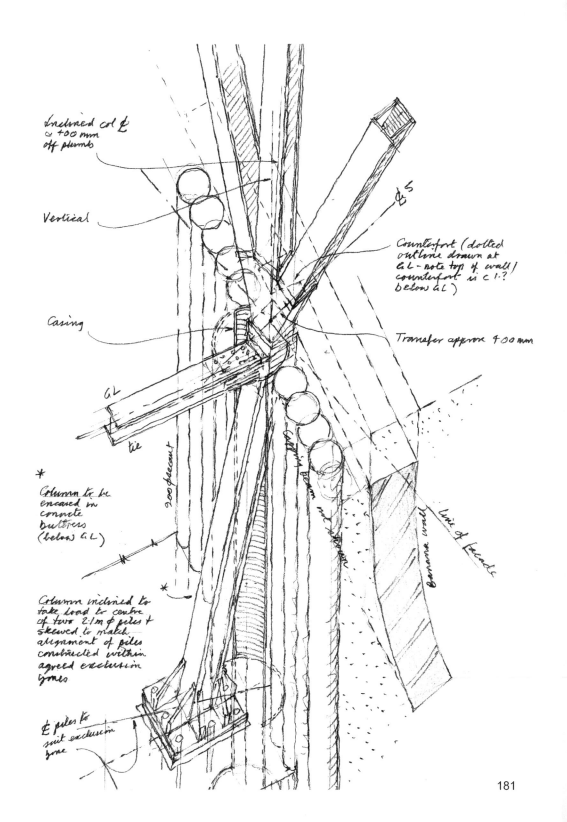

Inclined col ₵ ≈ 400 mm off plumb

Vertical

Casing

Counterfort (dotted outline drawn at GL - note top of wall/counterfort is c 1·? below GL)

Transfer approx 400 mm

GL

tie

200 precast

Carrying beam not shown

Banana wall

Line of façade

* Column to be encased in concrete buttress (below GL)

* Column inclined to take load to centre of two 2·1m ⌀ piles + skewed to match alignment of piles constructed within agreed exclusion zones

₵ piles to suit exclusion zone

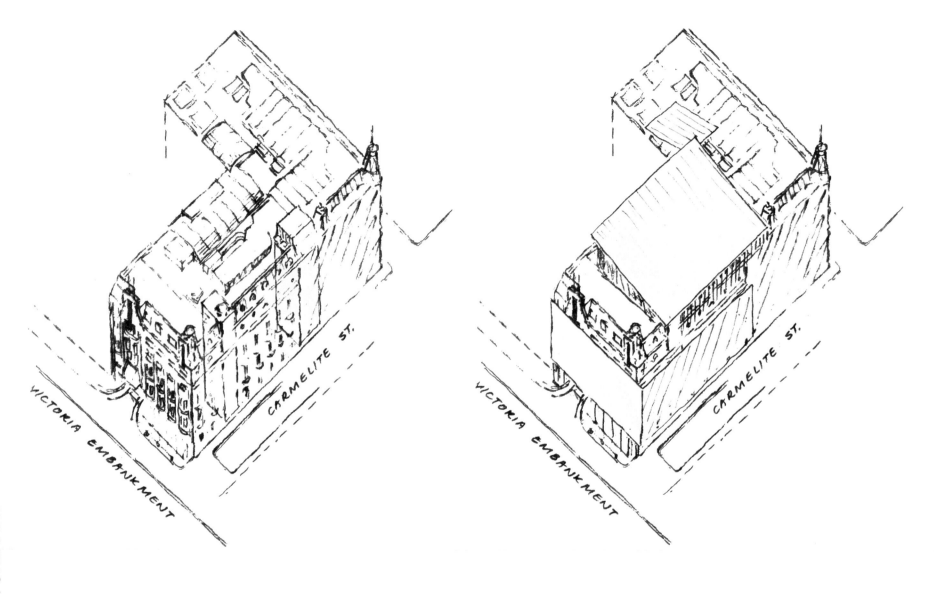

FIGURE 3.9/15

Carmelite, Victoria Embankment, London. The refurbishment of the Carmelite building on London's Victoria Embankment was a complex 'cut and carve' project – the building was re-clad, new accommodation provided at roof level and cores and staircases were re-configured. It's a good plan sometimes to use the same base drawing to produce a series of sketches, each showing a particular aspect of the works. The left hand sketch shows the location of existing stairs and the right hand sketch shows temporary weather protection to particular areas of the roof and the façade.

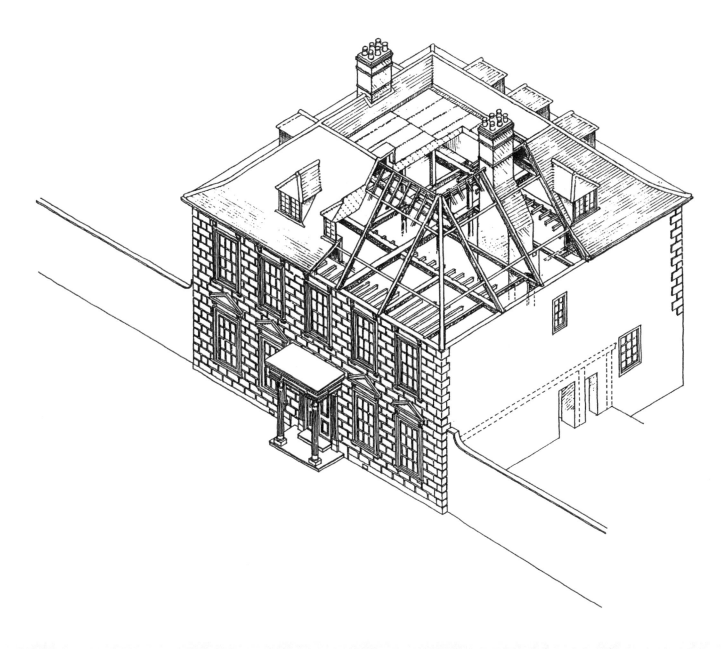

FIGURE 3.9/16

Turville Park Estate, London. Turville House is a handsome building in a beautiful setting. Major renovation and refurbishment needed to be carried out with a great deal of care and sensitivity. But it was the kind of project that was a joy to work on and producing explanatory drawings beyond the conventional was just a way of demonstrating commitment to a rewarding project and a wonderful client.

FIGURE 3.9/17
Tree Walk, Battersea Park, London. This is a long forgotten project constructed for the 1951 Festival of Britain and in a way is emulated by the much more recent tree walk at Kew Gardens. The Battersea Park walkway, dismantled many years ago, was designed as a series of trussed bridges slung from the trees rather than from separate columns; branches and trunks were checked individually as structural cantilevers.

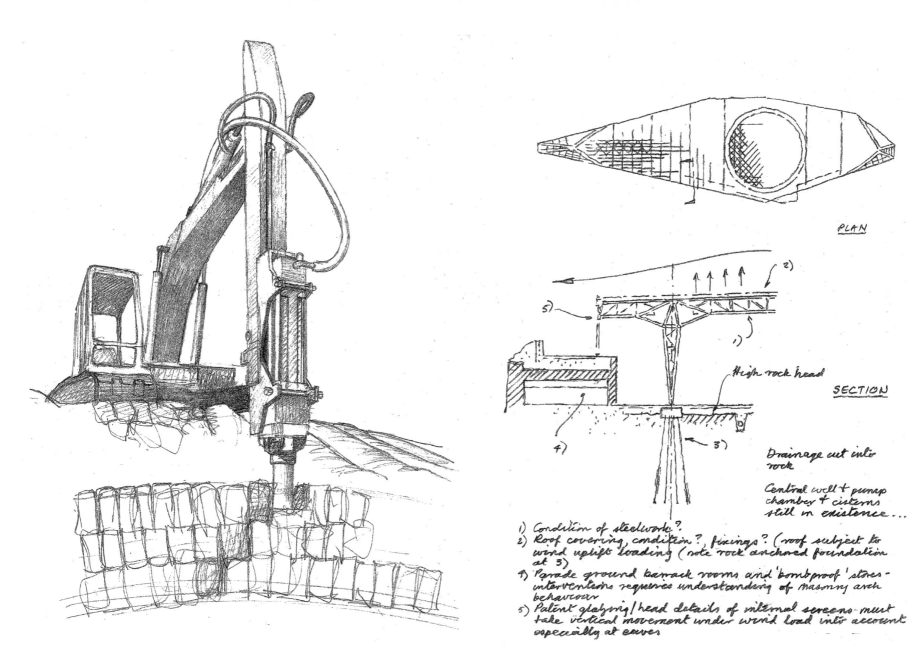

FIGURE 3.9/18

Fort Regent. Getting to know basic civil engineering plant is important. They are often great pieces of machinery to draw in their own right, but having an idea of the shape and size of excavators, piling rigs, mobile cranes and tower cranes is a first step towards knowing what is practical and buildable. The sketch on the left was drawn on site during the construction of the Fort Regent leisure centre in 1970; the sketches on the right are part of a bid document for a contract to refurbish the Fort.

PLAN

SECTION

High rock head

Drainage cut into rock

Central well + pump chamber + cisterns still in existence...

1) Condition of steelwork?
2) Roof covering, condition? fixings? (roof subject to wind uplift loading (note rock anchored foundation at 3)
4) Parade ground barrack rooms and 'bombproof' stores – interventions requires understanding of masonry arch behaviour
5) Patent glazing / head details of internal screens must take vertical movement under wind load into account especially at eaves

SKETCHBOOK

185

Piccadilly Estate, Piccadilly, London. Some projects, especially refurbishment or 'cut and carve' projects, are so complex you just have to learn your way around by drawing, and in some cases, like here at the Piccadilly Estate in London, by using 'cut-through' sections. By drawing, you find out what you don't know and then look for more detail or organise additional surveys. Another technique is to draw individual parts of an assembly of buildings and break down the proposals into bite-size pieces.

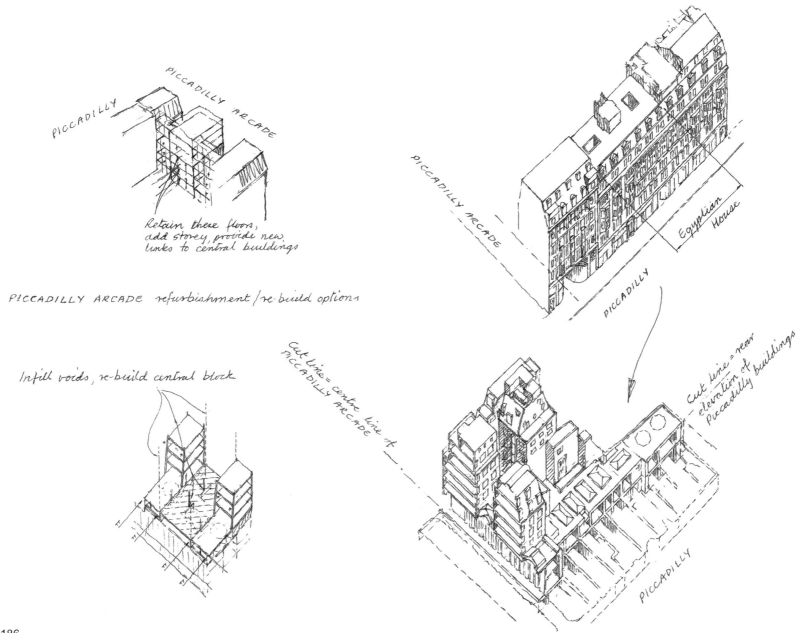

SKETCHBOOK

FIGURE 3.9/18

Fort Regent. Getting to know basic civil engineering plant is important. They are often great pieces of machinery to draw in their own right, but having an idea of the shape and size of excavators, piling rigs, mobile cranes and tower cranes is a first step towards knowing what is practical and buildable. The sketch on the left was drawn on site during the construction of the Fort Regent leisure centre in 1970; the sketches on the right are part of a bid document for a contract to refurbish the Fort.

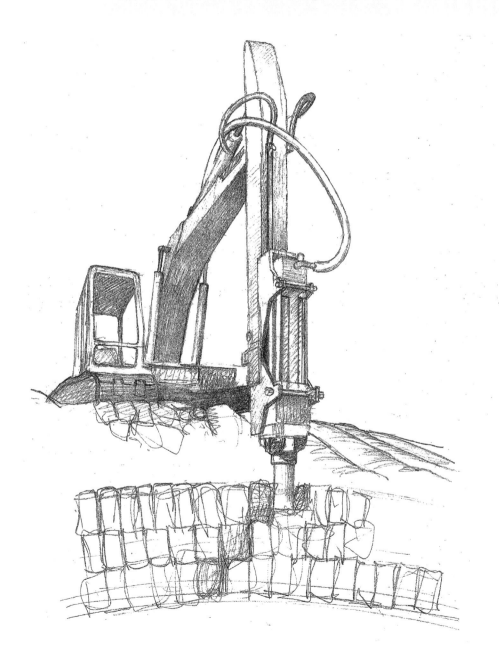

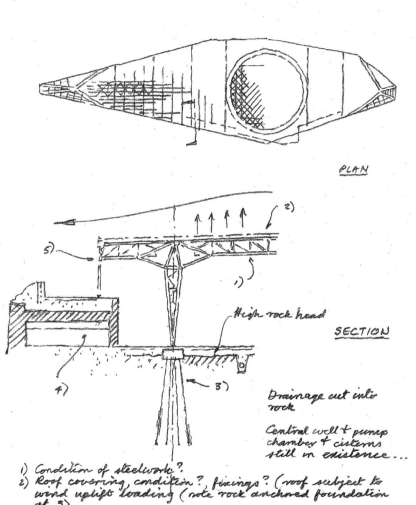

PLAN

High rock head

SECTION

Drainage cut into rock

Central well & pump chamber & cisterns still in existence...

1) Condition of steelwork?
2) Roof covering, condition?, fixings? (roof subject to wind uplift loading (note rock anchored foundation at 3)
4) Parade ground barrack rooms and 'bombproof' stores - intervention requires understanding of masonry arch behaviour
5) Patent glazing/head details of internal screens must take vertical movement under wind load into account especially at eaves

FIGURE 3.9/19

Piccadilly Estate, Piccadilly, London. Some projects, especially refurbishment or 'cut and carve' projects, are so complex you just have to learn your way around by drawing, and in some cases, like here at the Piccadilly Estate in London, by using 'cut-through' sections. By drawing, you find out what you don't know and then look for more detail or organise additional surveys. Another technique is to draw individual parts of an assembly of buildings and break down the proposals into bite-size pieces.

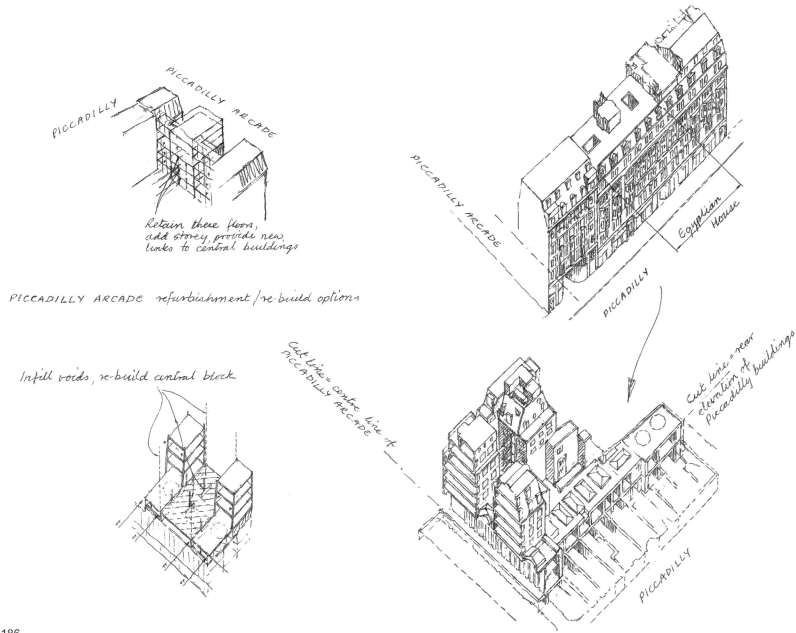

Retain these floors, add storey, provide new links to central buildings

PICCADILLY ARCADE refurbishment/re-build options

Infill voids, re-build central block

Cut line = centre line of PICCADILLY ARCADE

Cut line = rear elevation of Piccadilly buildings

FIGURE 3.9/20

Stanmore Court, St James's Street, London. Cut and carve refurbishment projects are challenging because engineers have to understand the existing building, the proposed alterations and temporary works needed to get from one to the other. At Stanmore Court temporary bracing in different forms was necessary to hold retained parts of the structure in place while major internal walls were demolished. Working closely with contractors to develop systems and sequences is essential.

STANMORE COURT ST. JAMES'S ST

WEST BLOCK TEMPORARY WALL   EAST BLOCK DEAD SHORES

SKETCHBOOK

187

3.10 LIGHTWEIGHT STRUCTURES

For the average structural engineer involved in the day-to-day design of buildings, lightweight structures are something of a challenge. They are easy to draw – ethereal, delicate, floating objects often with no visible means of support. But design isn't so easy and although we might use nature as our inspiration, it only takes us so far. For instance, trees can be beautiful lightweight things, but we all know they can flex hugely or even blow over in high winds.

In addition we are often working with unfamiliar material with different strength/weight relationships such as aluminium, fabrics, timber or even carbon fibre.

So how do we go about producing sketches that are going to lead to something that can be constructed in the real world? On one hand it's easy to draw something that just doesn't work, and on the other hand it's easy to design an object which is so obviously over-structured.

One good way is to look around you. Get a feel for existing lightweight structure: lamp posts, bus shelters, canopies, tents, awnings, railway gantries, pylons, road signs, exhibition stands, temporary stages, tables, chairs, stairs and so on. Many of these designs will have gone through a number of trial and error iterations or physical testing, so learn from the work that others have done.

Think of unusual structural forms such as monocoques, shells, space frames, membranes and tension structures, share your assumptions with others, sketch 'impressions', but keep practical design in mind.

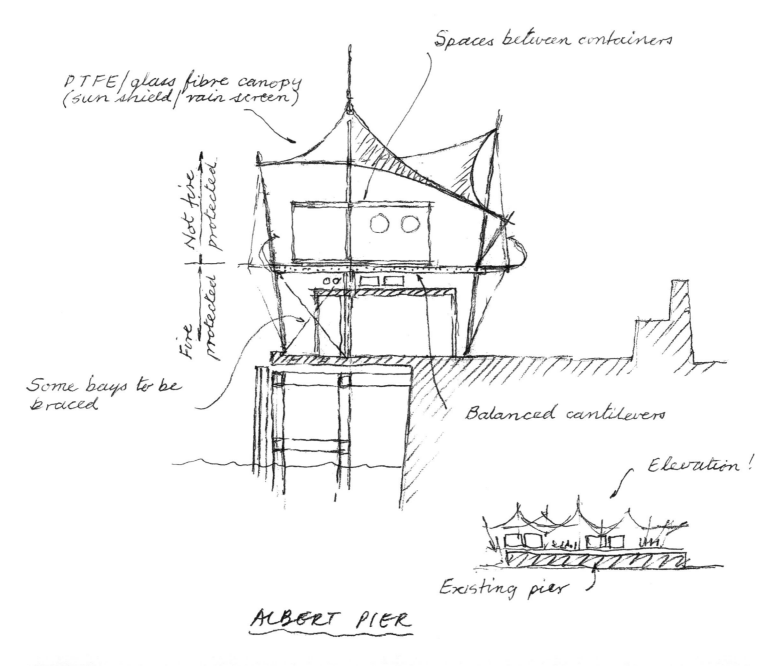

Spaces between containers

PTFE/glass fibre canopy
(sun shield/rain screen)

Not fire protected

Fire protected

Some bays to be braced

Balanced cantilevers

Elevation!

Existing pier

ALBERT PIER

FIGURE 3.10/1

Albert Pier, Jersey. A tiny scheme for an information centre on Albert Pier, Jersey – a number of concept ideas conveyed on quick 2D sketches, but these sketches are not spontaneous, they are not drawn from an architectural standpoint but they are the end result of lots of preliminary thoughts and scribbles on how things might fit together. Key features are lightweight structure, expressed stability system, off-site fabrication, sun screen/rain screen PTFE tensile roof and a defined fire protection strategy. The thumbnail sketch shows the overall visual impact the project might have on its surroundings.

Urban Oasis. Laurie Chetwood's Urban Oasis is a mobile sculpture designed to demonstrate some of the principles of sustainability. It is mobile in the sense that it has motorised arms and it is demountable and transportable. The exhibit provides a platform to demonstrate the use of PVs (photo-voltaics), rainwater harvesting, wind generated electrical power and the use of fuel cells. Structural design was based on a five-way cable-supported mast and on folding arms, supporting petals covered with PVs, deploying in calm, sunny conditions but retracting out of harm's way in high winds.

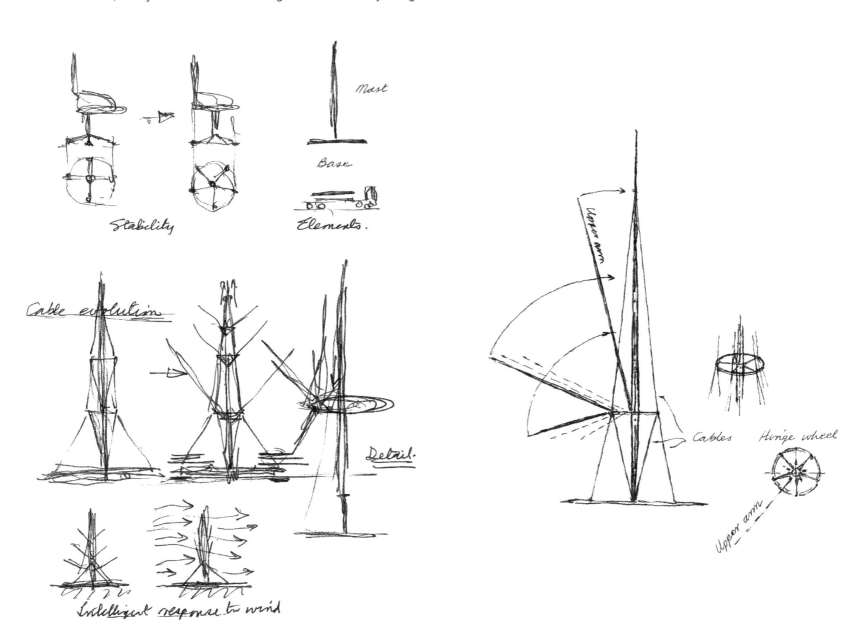

SKETCHBOOK

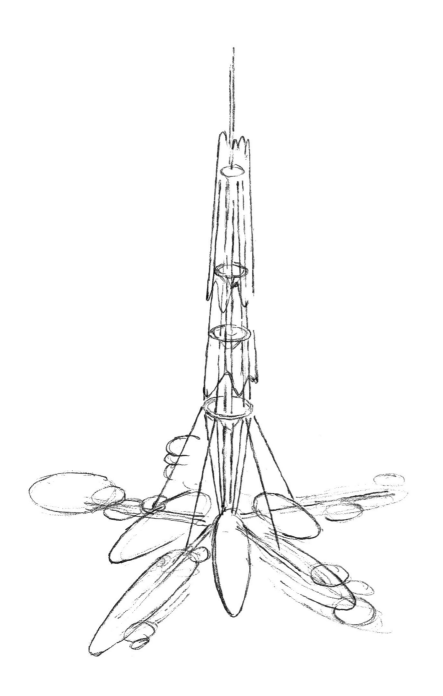

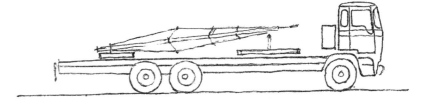

FIGURE 3.10/3

Urban Oasis. Urban Oasis was designed as a stiffened mast tied down to kentledge by a separate array of cables. The architecture is very much in keeping with other works by Laurie Chetwood, including his Butterfly House and the Perfumed Garden, all of which are based on lightweight structures. In fact the mast here was reused in the Perfumed Garden (q.v.) and both it and Urban Oasis were centrepieces in gold medal winning gardens at the Chelsea Flower Show.

FIGURE 3.10/4

Spine designs. Two designs based loosely on biological references to spines, vertebrae, transverse processes and ligaments and resolved into struts and tendons. The canopy to the upper left was a design for a shopping mall, while the stair lower left is a staircase design for an architect's office in Central London. The stair was constructed as a pair of trussed beams and the beams themselves are aluminium yacht masts.

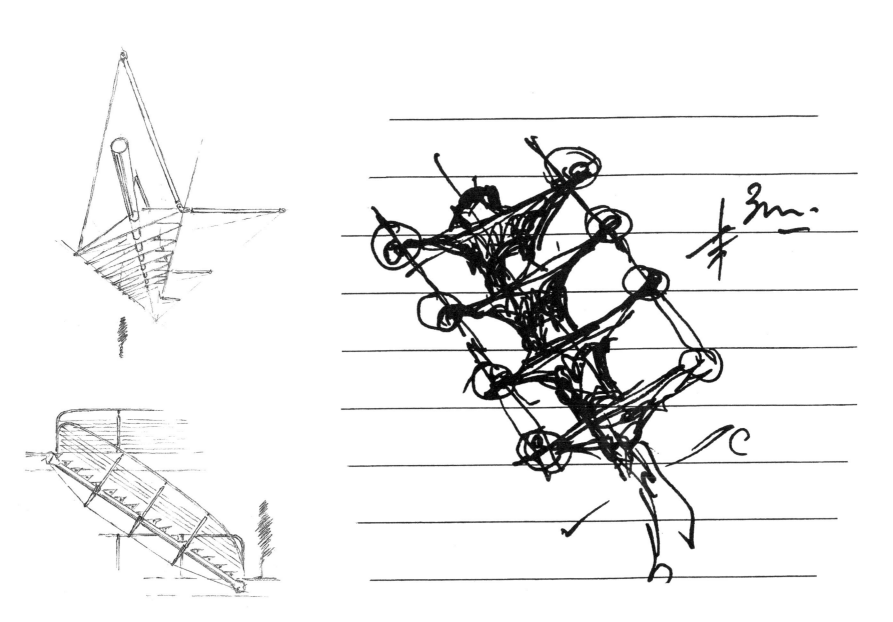

FIGURE 3.10/5

Covered arcade for Sainbury's at Richmond. The quest for something different usually means looking at seemingly endless options, and sometimes this is not very rewarding when a design doesn't emerge through logical progression. Best to record the ongoing search with a series of simple two-dimensional elevations and sections.

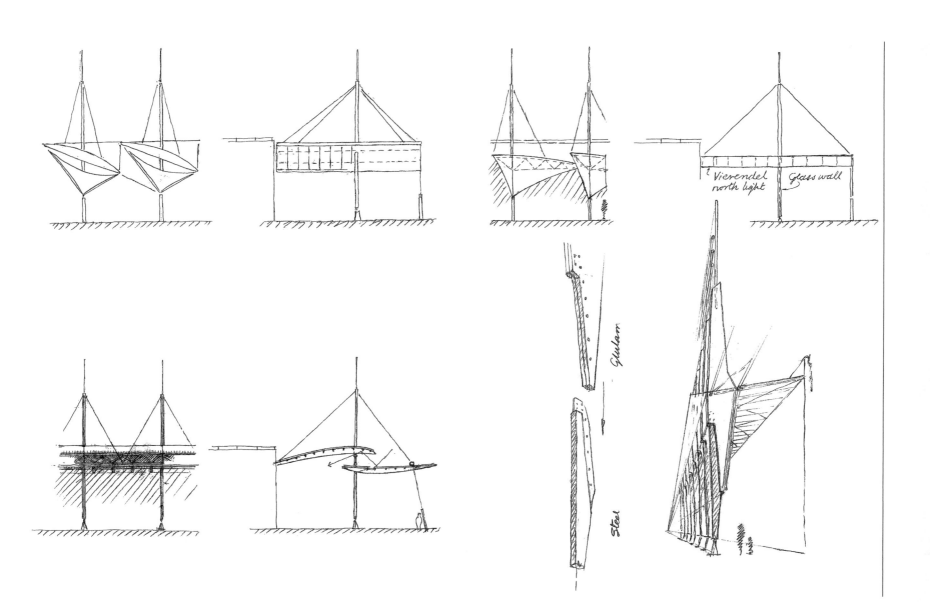

FIGURE 3.10/6

Perfumed Garden, Chelsea Flower Show, London. Laurie Chetwood's perfume-themed garden at the Chelsea Flower Show was another gold medal winning design, awarded for the garden of course not the structure. The centrepiece was built around the mast recycled from his earlier exhibit but this time wrapped in cables supporting PTFE fabric spirals. The sketch summarises the design principles where struts from the mast and the cables provide outlines for the spirals and fine adjustment to the spirals is achieved using adjustable ties onto the cables.

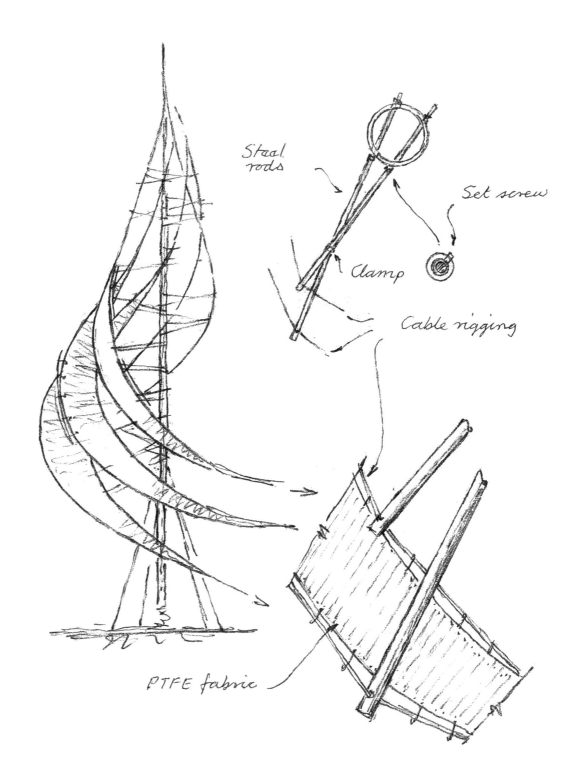

Recycled mast (from 'Urban Oasis')

Steel rods

Set screw

Clamp

Cable rigging

PTFE fabric

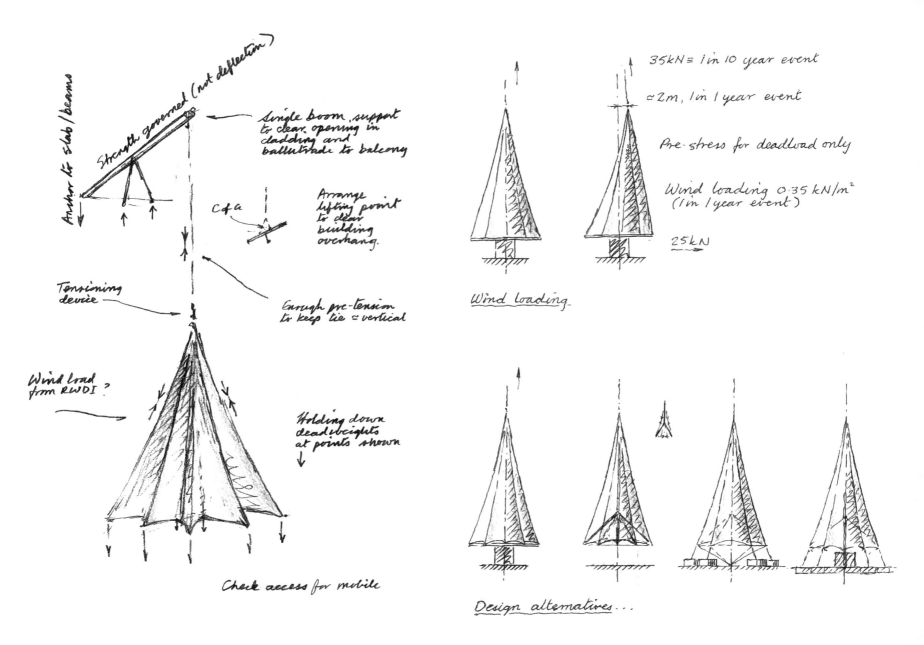

Strength governed (not deflection)

Single boom support to clear opening in cladding and ballutrade to balcony

Anchor to slab/beams

C of a

Arrange lifting point to clear building overhang.

Tensioning device

Enough pre-tension to keep tie ≃ vertical

Wind load from RWDI?

Holding down dead weights at points shown

Check access for mobile

35kN ≡ 1 in 10 year event

≃2m, 1 in 1 year event

Pre-stress for deadload only

Wind loading 0.35 kN/m² (1 in 1 year event)

25kN

Wind loading

Design alternatives...

FIGURE 3.10/7

Christmas Tree Project, Merchant Square, Paddington Basin, London. Not everyone's idea of a Christmas tree, but this unconventional design was chosen by a developer to adorn his construction site at Merchant Square. The concept relied on suspending a tree-shaped tensile structure from an upper floor of a partially completed building. A number of options were developed for the base of the tree; all had to take account of the fact that lightweight structures are potentially susceptible to wind damage.

FIGURE 3.10/8

Petrol filling station, Sainsbury's, Watford. Only a petrol filling station but the design team aim was simplicity and simplicity is not achieved without a great deal of effort. Here after much to-ing and fro-ing between architect and engineer, the repetitive nature of the final touching canopies scheme is illustrated by a single point perspective. Sketch reproduced by kind permission of Sainsbury's Supermarkets Ltd.

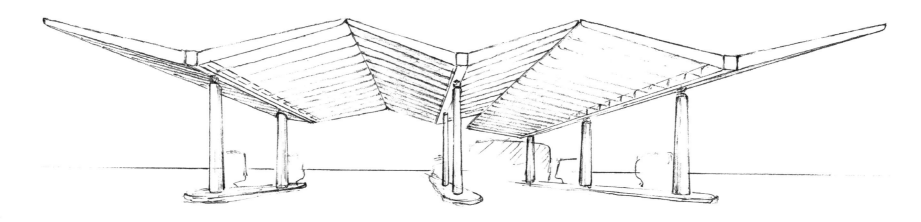

3.11   UNUSUAL BUILDING TYPES

It is difficult to define unusual building types but just occasionally you come across something that doesn't fit into an easily recognised category. The aquarium at Silvertown, an observation tower in Dubai, a mosque in Abu Dhabi and signage supports at a well-known bookshop in London are all projects which for me have been different and 'one offs'.

The best example I have come across is the proposed 'Biota!' world class aquarium at Silvertown in Tower Hamlets which sadly was one of the first victims of the 2007/8 economic downturn. We were designing curvilinear tanks with 300mm thick polycarbonate windows, air supported polyethylene pillows and hybrid timber/steel gridshell roofs, all with a biodiversity and 'conservation of the seas' theme in mind – what a shame it was never built, at least not yet!

It is, however, next to impossible to come to any general conclusion with regard to conceptualising or sketching relevant structures – just enjoy the difference.

SKETCHBOOK

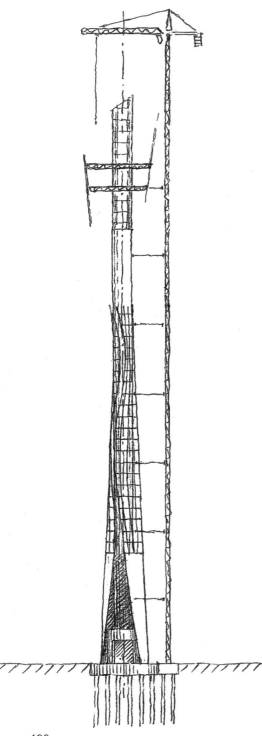

FIGURE 3.11/1
Observation tower, Middle East. A very distinctive and beautiful proposal for an observation tower in the Middle East which, sadly, was never developed beyond the initial concept ideas. The design was based on a slip-formed circular core with an outer metal clad steel skeleton – slenderness would have been a major design issue and the tower would have required some form of damper to control wind induced oscillation.

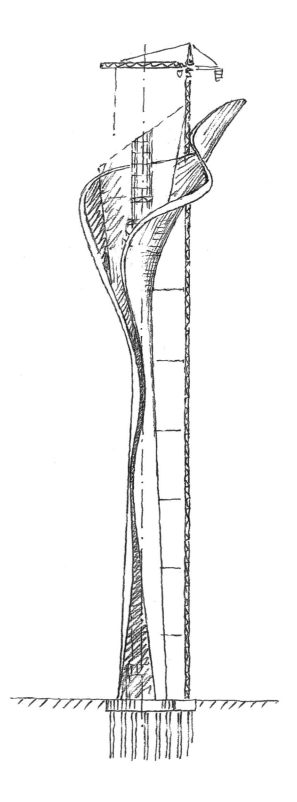

FIGURE 3.11/2

Mosque, Middle East. The concept for the structure of the mosque was unusual if not unique. Permanent columns are erected and braced to form a central stable tower. Perimeter curved members are then positioned around the tower and infilled with secondary steels and rebar mesh ready for the application of sprayed concrete. Once complete, the membrane action of the shell surface provides the stability and bracing to the internal columns.

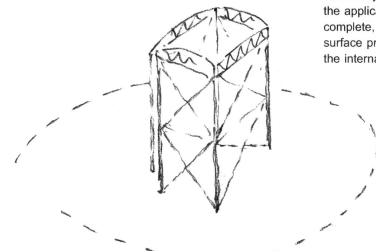

Stage 1   Temporary central tower

Stage 2   Primary frame

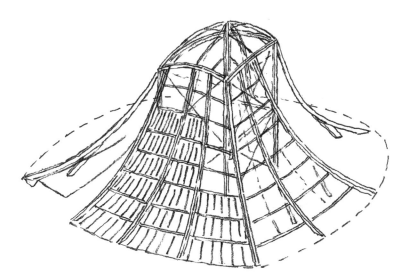

Stage 3   Secondary framing

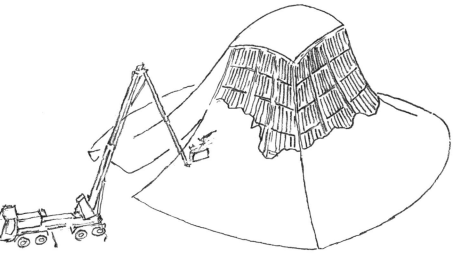

Stage 4   sprayed concrete

DS3, Canary Wharf, London. A column in this building and structure above had to be lifted vertically by 250mm. First the structure above was supported on a temporary 'A' frame, concrete around the base of the column was broken out and then a hydraulic jack could be inserted. The existing structure below the jack was strengthened to resist forces that would be generated by the lifting process. This cutaway drawing was used in a report explaining the modus operandi.

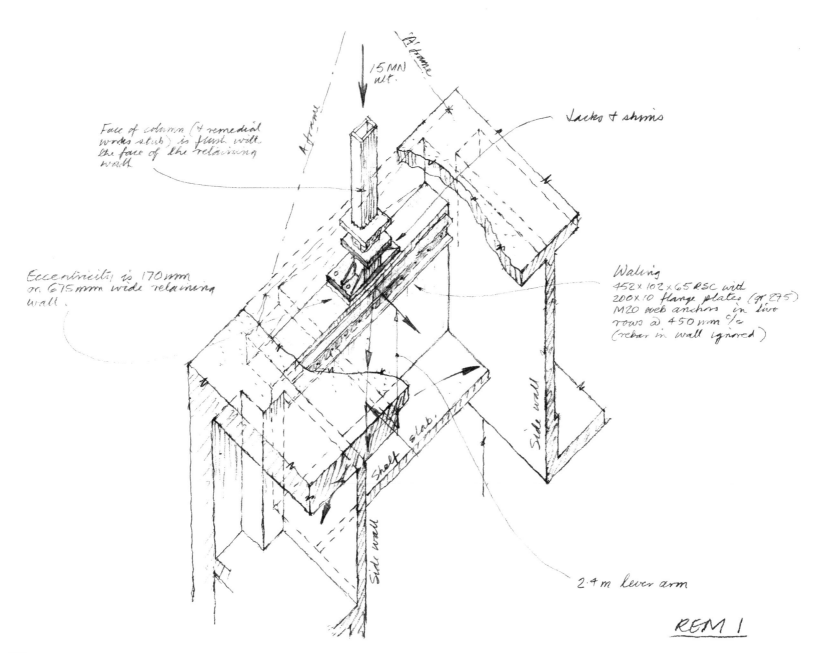

SKETCHBOOK

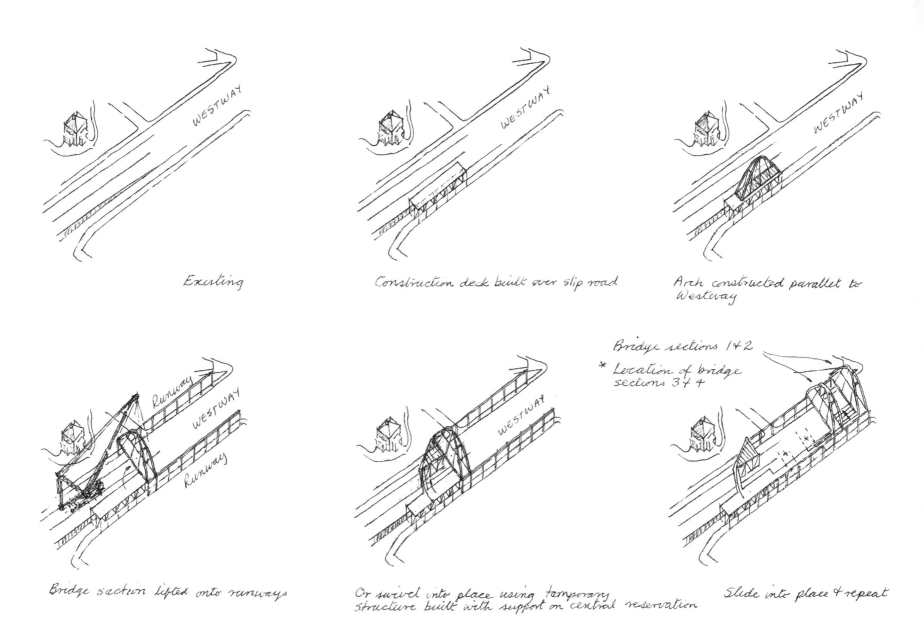

Existing

Construction deck built over slip road

Arch constructed parallel to Westway

Bridge section lifted onto runways

Or swivel into place using temporary structure built with support on central reservation

Slide into place & repeat

Bridge sections 1 & 2

* Location of bridge sections 3 & 4

FIGURE 3.11/4

Green park over Westway, London. The ancient borough of Paddington has long been cut in two by the A40 Westway. Sir Terry Farrell's bold scheme to reconnect Paddington Green on the north side of the highway with Paddington Basin on the south, involved the construction of a lengthy pedestrianised green bridge over six lanes of roadway. The engineering solution was based on a series of arched bridges, assembled on a platform built over a slip road. Each bridge would then be lifted into position during a temporary road closure, and then slid to its final resting place along parallel runways built either side of the highway.

3.12   BIOLOGICAL REFERENCES

In engineering and architecture, a structure is a 'body or assemblage of bodies in space to form a system capable of supporting loads'. This simple definition applies equally well to physical structures in the natural world of trees, skeletons, anthills, beaver dams and salt domes. In biology, structures exist at all levels of organisation, ranging hierarchically from the atomic and molecular to the cellular, tissue, organ, organismic, population and ecosystem level.

Therefore it is not surprising that engineers and architects have used references to the natural world for inspiration and in some cases for retrospective justification.

But as per physical models, scale effect presents a sometimes insurmountable problem. Linear relationships between size, structural behaviour and applied loadings do not exist and therefore it is impossible to scale up, say, the structure of a flower head and expect it to work as a helicopter landing pad.

On the other hand, there are useful analogies that can be used to build a design solution. Struts and ties versus spinal columns and tendons, light but strong tubular structures, diaphragms and shells, natural arches and domes, cable nets and spider webs, and so on.

Quick sketches of reference material are often convincing, even if retrospective, and certainly give the design process an added dimension.

SKETCHBOOK

FIGURE 3.12/1

Royal Albert Bridge, Saltash, Cornwall. Someone, some time ago, made an interesting comparison between the workings of the Royal Albert Bridge at Saltash and the diplodocus. The diplodocus is the longest if not the biggest sauropod known from a complete skeleton and for such a huge beast, it is reasonable to assume that compression in its arched spine was counterbalanced by tension in its underbelly, just like the interaction between Brunel's arch and the suspension system. Brunel died before the diplodocus skeletons were discovered, so sadly there is no possibility that Mr Brunel ever noted a similarity.

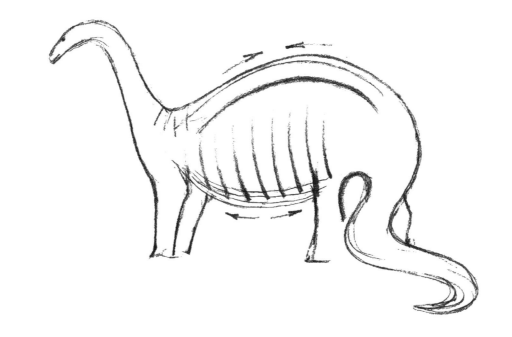

SKETCHBOOK

203

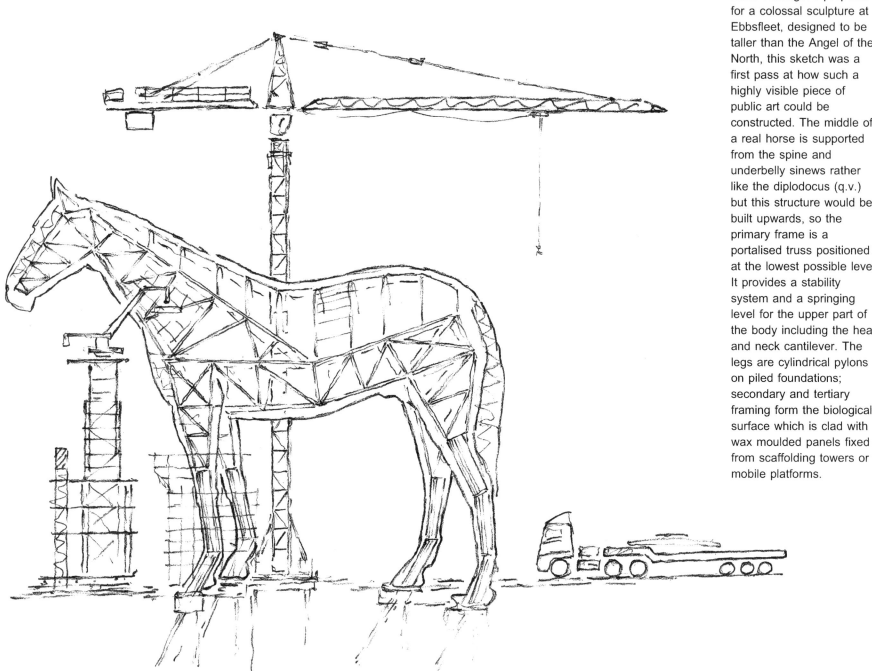

FIGURE 3.12/2
Ebbsfleet design competition. Inspired by Mark Wallinger's proposal for a colossal sculpture at Ebbsfleet, designed to be taller than the Angel of the North, this sketch was a first pass at how such a highly visible piece of public art could be constructed. The middle of a real horse is supported from the spine and underbelly sinews rather like the diplodocus (q.v.) but this structure would be built upwards, so the primary frame is a portalised truss positioned at the lowest possible level. It provides a stability system and a springing level for the upper part of the body including the head and neck cantilever. The legs are cylindrical pylons on piled foundations; secondary and tertiary framing form the biological surface which is clad with wax moulded panels fixed from scaffolding towers or mobile platforms.

FIGURE 3.12/3

Staff accommodation, Waterloo Station, London. Post-rationalised perhaps, but this was a way of showing how the column structures support the staff accommodation raft at Waterloo Station (q.v.).

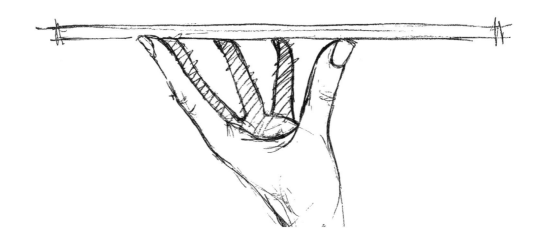

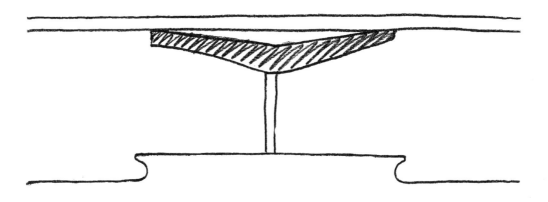

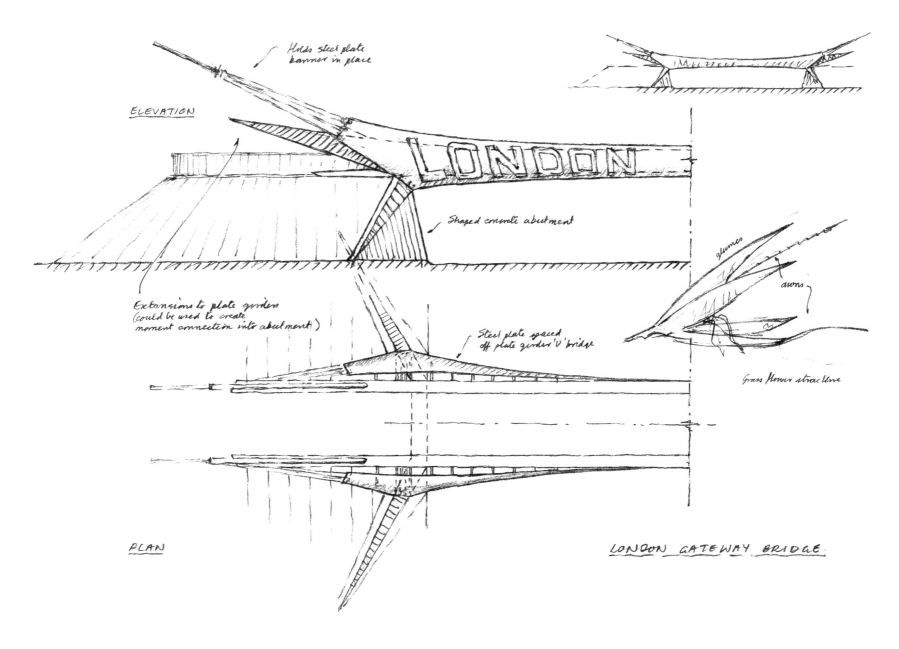

ELEVATION

Holds steel plate banner in place

LONDON

Shaped concrete abutment

Extensions to plate girders (could be used to create moment connection into abutment)

Steel plate spaced off plate girder 'U' bridge

glumes

awns

Grass Flower structure

PLAN

LONDON GATEWAY BRIDGE.

FIGURE 3.12/4

London Gateway Bridge. Nothing more than a dressed up single span plate girder 'U frame' bridge. (A 'U frame' bridge is a reference to the cross-section of the structure where cantilever action from the bridge deck provides sufficient restraint to the otherwise unrestrained top flanges of the parapet beams.) Thin steel banner sheets are held in place by struts pointing away from the road. The inspiration for the cluster of shapes at each abutment comes from the awns and glumes of a typical grass flower.

3.13   DETAIL

Detail design is perhaps less exciting than whole building design or master planning work but is often the key to successfully building out a project. More often than not, 2D plans and elevations of detail are produced as essential standard practice but 3D sketches are a helpful addition because they provide almost instantaneous appreciation and understanding. It has to be said, however, that 3D images are usually not good enough on their own – and, in any case, are drawn when the design has been developed from 2D drawings.

In structural engineering, detail is particularly important with regard to the design of connections. Connections are critical in terms of transferring load from one member to another, especially where joining members may be made from different materials working at different allowable or ultimate stress limitations. In addition, connections may have to allow for rotation or provide different degrees of freedom with regard to movement. It is a well known fact that getting load from one structural member to another is not always easy.

Sketching detail helps produce practical solutions and helps show how things fit together. It is a way of solving problems as well as an essential way of developing and communicating solutions. Architects and designers love connections and quite often embellish and express them in their compositions rather than have them hidden away.

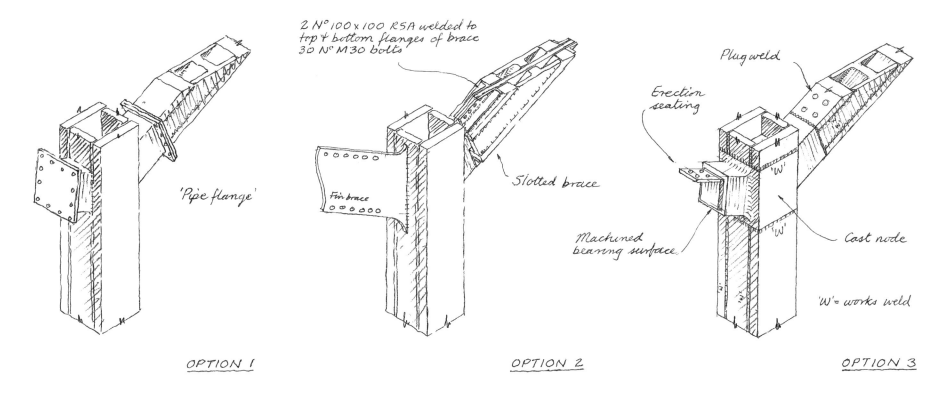

2 N° 100 × 100 RSA welded to
top & bottom flanges of brace
30 N° M30 bolts

Plug weld

Erection
seating

'Pipe flange'

Fin brace

Slotted brace

Machined
bearing surface

Cast node

'W'

'W'

'W'= works weld

OPTION 1

OPTION 2

OPTION 3

FIGURE 3.13/1

Northgate, London. There are many ways of putting major steelwork connections together but in the UK this is quite often left to the fabricator. However, at Northgate, where major steelwork was to be on view at highly visible junctions between struts and columns, aesthetic quality and constructability had to be studied from the beginning by the architect and engineer working together.

FIGURE 3.13/2

1 Blackfriars Road, London. Hybrid structural frames consisting of part reinforced concrete and part structural steel are not common in high-rise construction in the UK. Sometimes however, structural steel is introduced in the lower storeys of a largely concrete residential tower, to deal with transfer spans over lobby areas or to suit column shapes and sizes at ground level. When hybrid structures of this type are proposed, special attention must be given to the interface between the two frame types, especially with regard to buildability and local bearing stress.

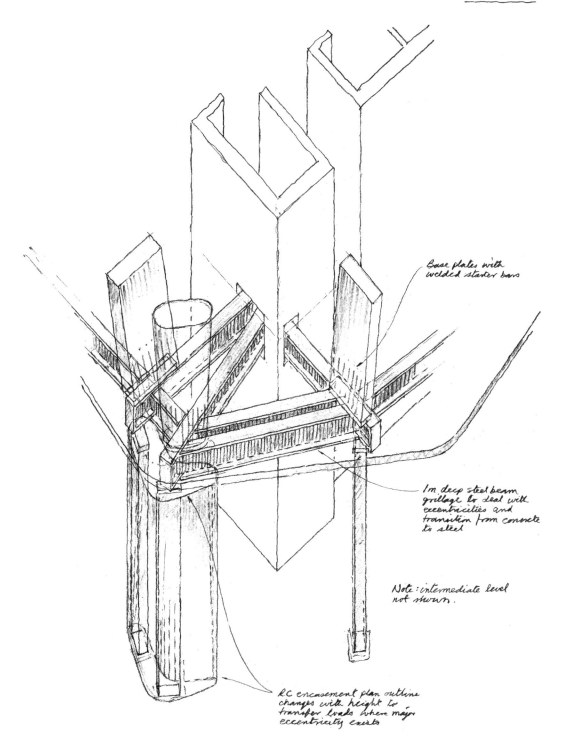

Base plates with welded starter bars

1m deep steel beam grillage to deal with eccentricities and transition from concrete to steel

Note: intermediate level not shown.

RC encasement plan outline changes with height to transfer loads where major eccentricity exists

The Wellcome Trust HQ, Euston Road, London. The southern entrance to the Euston Road Undergound station is located in the northwest corner of the Wellcome Trust's headquarters building. Keeping access and egress to the railway open throughout the construction period involved switching people from the old route to a temporary route and then back to the original albeit slightly widened and reconfigured route. This sketch shows the temporary stair and how it was positioned in relation to piling and pile cap works required to support the new building.

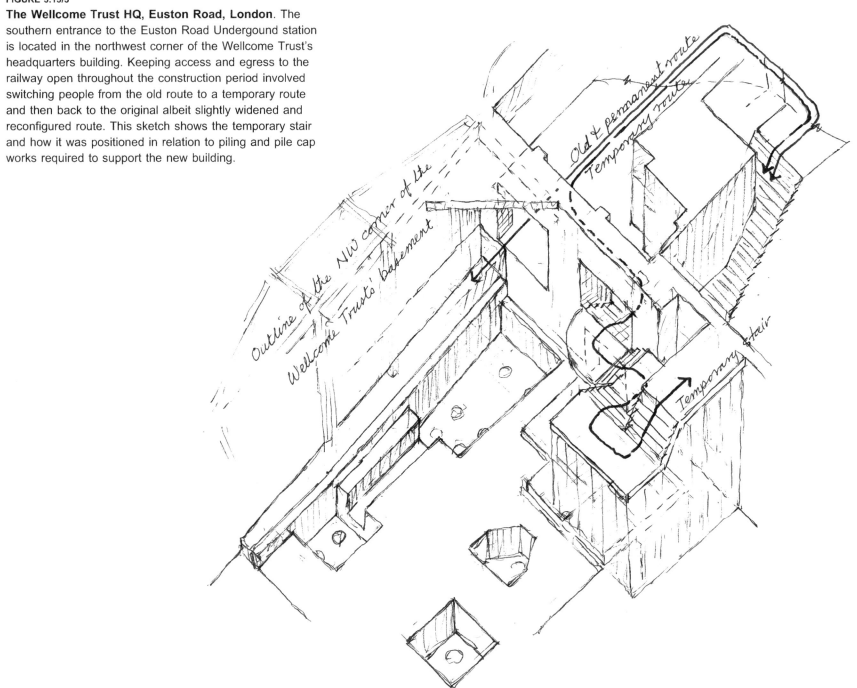

FIGURE 3.13/4
European HQ, Brentford, London. A scheme for a European headquarters at Brentford was developed by Terry Farrell & Partners. It was based on a concrete frame with thin slab edges to allow maximum natural light into the floors. Nowadays most unitised cladding systems are fixed using manipulators working off the completed slabs and not cassette systems which take too much crane hook time. The site was eventually developed for Glaxo Smith Kline.

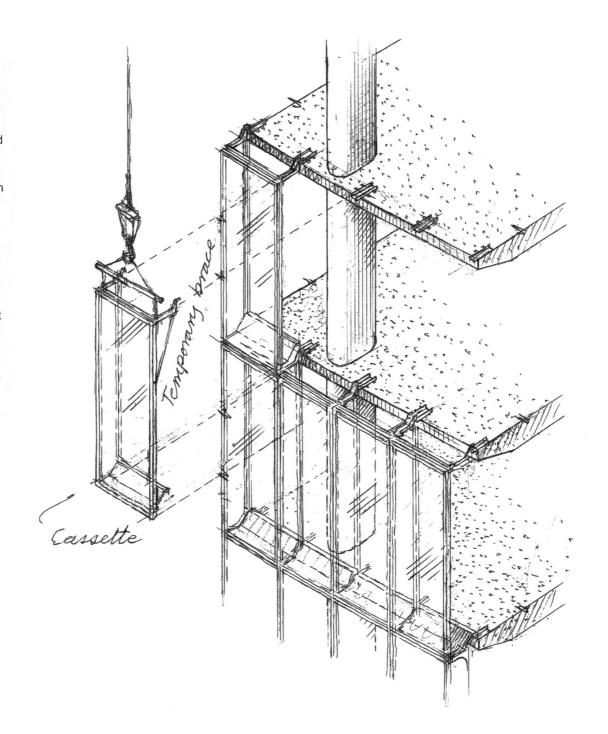

Temporary brace

Cassette

SKETCHBOOK

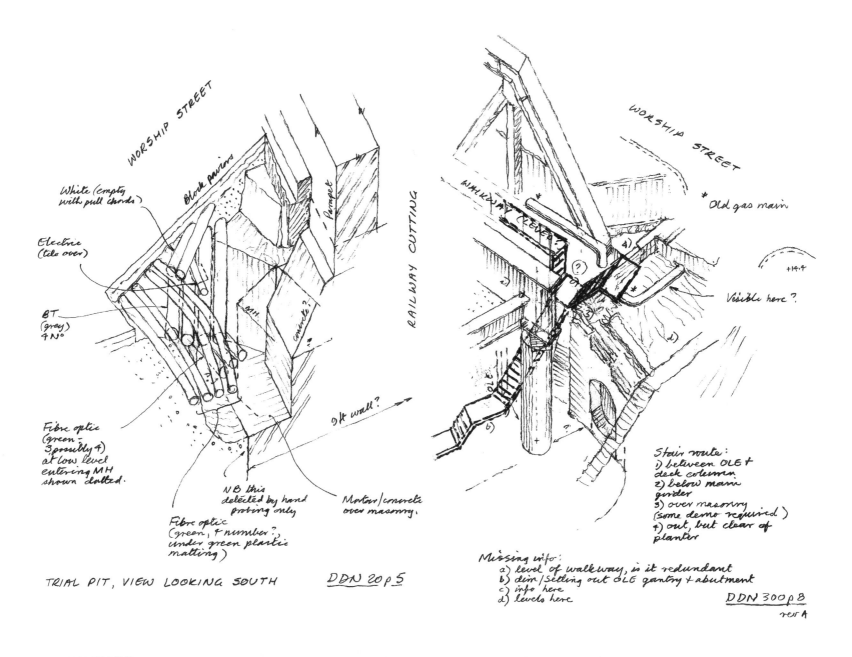

WORSHIP STREET

Block pavoirs

White (empty
with pull chords)

Electric
(tile over)

BT
(grey)
4 No

Fibre optic
(green -
3 possibly 4)
at low level
entering MH
shown dotted.

Parapet

concrete ?

RAILWAY CUTTING

9 ft wall?

NB this
detected by hand
probing only

Fibre optic
(green, 4 number ?,
under green plastic
matting)

Mortar/concrete
over masonry.

TRIAL PIT, VIEW LOOKING SOUTH   DDN 20 p 5

WALKWAY LEVEL

WORSHIP STREET

* Old gas main

+14.4

Visible here ?

Stair route:
1) between OLE +
deck column
2) below main
girder
3) over masonry
(some demo required)
4) out, but clear of
planter

Missing info:
a) level of walkway, is it redundant
b) dim/setting out OLE gantry + abutment
c) info here
d) levels here

DDN 300 p 8
rev A

FIGURE 3.13/5

Norton Folgate, London. Perhaps the sight of congested services exposed in a trial pit is not the most inspiring thing from an engineering or a sketching point of view, but it is really surprising how a sketch like this can bring greater clarity than photography. Shadows are removed and the line weight is chosen to separate the important from the unimportant. The right hand sketch is the same location and is part of a study to plan an escape stair from the railway to street level. One great advantage of an orthogonal sketch like this over a perspective, is that it can be (and is) drawn to scale, both in plan and vertically.

212

3.14   MASTER PLANNING

Master planning work ranges from the large-scale urban design of cities through to the more detailed master planning of complex individual sites, although usually it is the big picture stuff which takes account of sustainable and humane environments and is respectful of the climate, context and history of the site.

In the United States it is known as comprehensive planning, a term employed by land use planners to describe a process that determines community goals and aspirations in relation to community development. The outcome of comprehensive planning is the 'comprehensive plan', which dictates public policy in terms of transportation, utilities, land use, recreation, and housing. Comprehensive plans typically encompass large geographical areas, a broad range of topics, and cover a long-term time horizon.

In Canada, comprehensive planning is generally known as strategic planning or visioning. It is usually accompanied by public consultation. When cities and municipalities engage in comprehensive planning the resulting document is known as an official community plan.

In London, Sir Terry Farrell is one of the best known master planning architects, with a deep knowledge of the city and a persuasive analytical approach whether based on industrial development along the tributaries of the Thames, thoroughfares through London or the effect that Victorian railways and canals had on the development of the conurbation.

Contributing to the master planning process is about rationalising information, explaining historical data and helping to describe new strategies for long term aspirations. It is often as much to do with 'what not to sketch' rather than 'what to sketch' – there is a danger of showing too much detail and losing the overall message.

FIGURE 3.14/1

Ludgate and Sampson House. This is no more than an early stage overview showing the scale of the development on London's South Bank: the massing and the impact the railway would have on demolition, logistics, vibration isolation, phasing, interface with the railway, adjacent construction sites etc. It was produced as a base sketch to be separately annotated to explain the engineering approach to many of these issues.

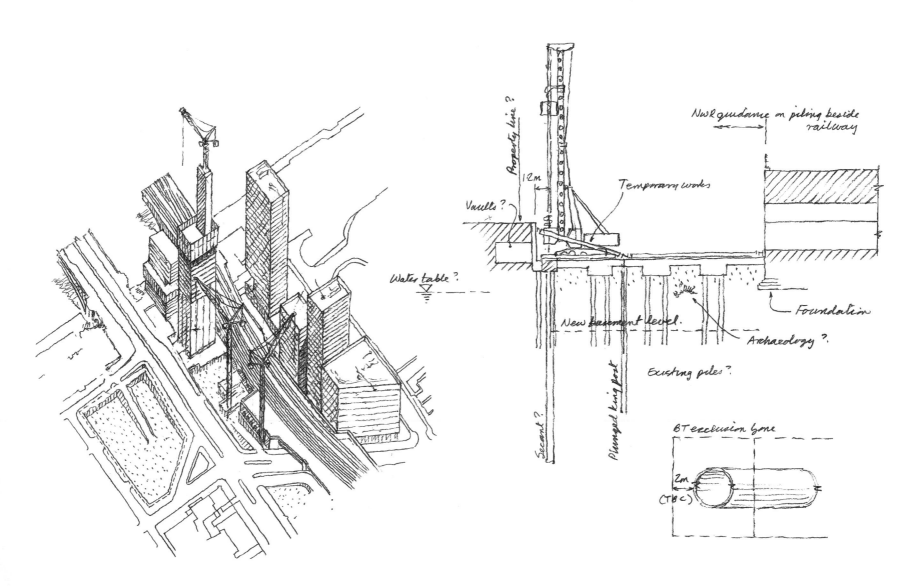

3.15 BASEMENTS AND TUNNELS

Basements

Basement construction on unconfined sites is relatively straightforward. Given reasonably good geotechnical conditions and no groundwater, adequate perimeter space and provided the basement is no more than one or two levels deep, the solution will nearly always involve 'open-cut'. The sides of the excavation are battered back, the foundation and substructure built and then the perimeter is backfilled against newly constructed retaining walls. But in the city centre, things are rarely that easy – construction is much more challenging on a heavily constrained urban site where perimeter access from outside of the site is impossible but where every square inch of developable land is valuable.

To illustrate suitable methodologies and to draw possible construction sequences it is important to have a basic understanding of available techniques and limitations of plant. Nearly all sites in non-rock geology will employ some sort of embedded wall solution. An embedded wall is a wall constructed so that to one extent or another it can act as a cantilever. Types include:

• Sheet pile walls: Interlocking steel sheet pile walls are driven into the ground, or more commonly in urban environments, hydraulically pressed or jacked into the ground to avoid unnecessary noise and vibration.
• Secant piled walls: Unreinforced female piles are constructed at given centres and then harder reinforced (and deeper) piles are constructed between, and cut into, the female piles to give a continuous 'leak proof' wall.

• Contiguous piled walls: Reinforced piles constructed adjacent to each other – usually only suitable where ground water is not a big problem.
• Diaphragm walls: Panels cast in a trench first held open by using thixotropic material (bentonite) introduced into the trench as excavation proceeds. Reinforcement cages are lowered through the bentonite which is then displaced by pumping concrete, via tremie tubes, to the base of the panel.
• King post walls: Here steel posts are forced into the wet concrete of newly formed piles. The piles and king posts are installed at 4 or 5m centres and then reinforced concrete infill walls are cast 'top-down' between the king posts as excavation proceeds.

These walls, which may eventually become part of the basement box, will not always be capable of acting as true cantilevers over the depth of the basement and will require some form of temporary propping as the excavation is taken deeper. Propping sequences are therefore important. For very deep basements, the permanent structure is often built 'top-down' to provide lateral support to the perimeter walls.

Top-down construction takes place from ground level downwards, rather than excavation to the bottom of the basement and working up. At The Place, 500 × 500mm steel columns were lowered or 'plunged' from ground level through empty pile bores into freshly poured concrete. The ground level slab was then cast on grade. Once it had gained strength the slab was capable of propping the perimeter embedded walls and could be supported on the plunge columns, allowing excavation to commence below the slab.

Space requirements for plant, how closely it can operate against a vertical boundary, access to site and acceptable propping arrangements must all be taken into account.

Tunnels

Tunnel construction is even more specialised than basement construction – in general, methodologies rely on ground type and tunnel depth. Some very sophisticated techniques are used for shallow tunnelling and use a combination of jacking 'pipes' ahead of the tunnel face and grouting.

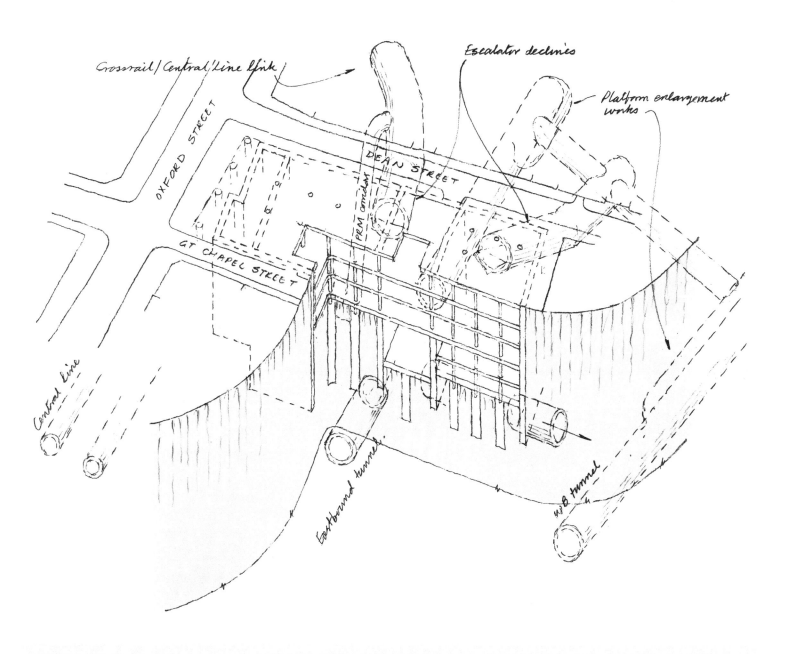

Crossrail/Central Line link

Escalator declines

Platform enlargement works

OXFORD STREET

DEAN STREET

PRM corridor

GT CHAPEL STREET

Central Line

Eastbound tunnel.

WB tunnel

FIGURE 3.15/1

Over-site development, Oxford Street, London. Over-site developments on the Crossrail route provided challenges and opportunities in equal measure. This cut-away section showed a proposed development over the western ticket hall of Crossrail's station at Tottenham Court Road. Understanding where the various tunnels would be constructed was vital with regard to founding the new building and predicting ground movement that might affect either the building if it were built first or the tunnels if they were constructed first.

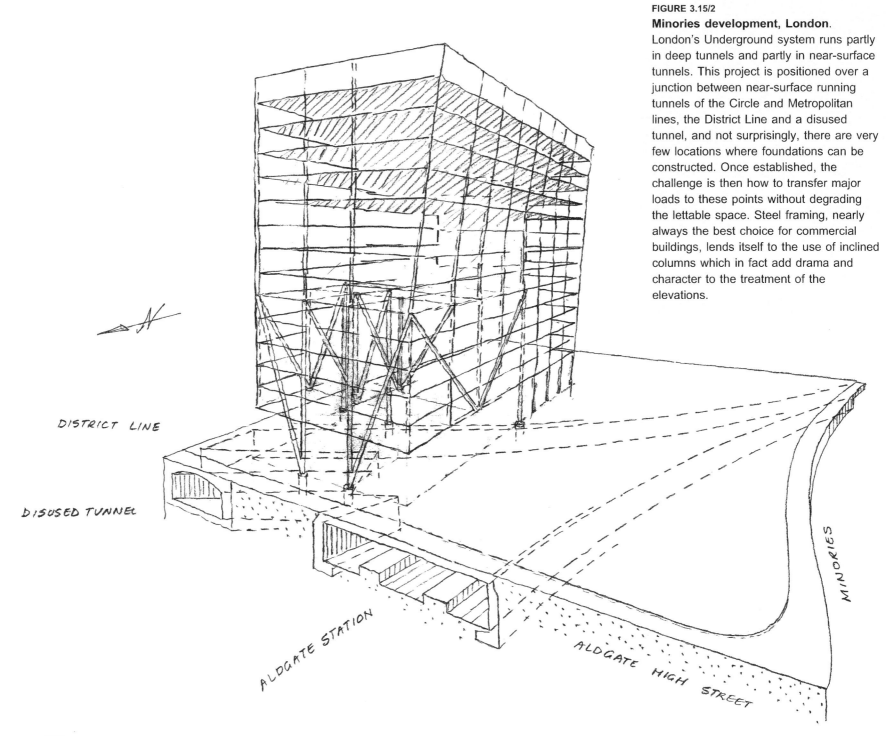

FIGURE 3.15/2

Minories development, London.

London's Underground system runs partly in deep tunnels and partly in near-surface tunnels. This project is positioned over a junction between near-surface running tunnels of the Circle and Metropolitan lines, the District Line and a disused tunnel, and not surprisingly, there are very few locations where foundations can be constructed. Once established, the challenge is then how to transfer major loads to these points without degrading the lettable space. Steel framing, nearly always the best choice for commercial buildings, lends itself to the use of inclined columns which in fact add drama and character to the treatment of the elevations.

DISTRICT LINE

DISUSED TUNNEL

ALDGATE STATION

ALDGATE HIGH STREET

MINORIES

FIGURE 3.15/3

Kingsgate House, Victoria Street, London. Masonry tunnels are very susceptible to damage caused by ground movement. Here a deep basement is to be constructed beside the Circle and District line's cut-and-cover tunnel running parallel to Victoria Street in London. Although some ground movement is inevitable, construction sequence must be designed to limit movement to the absolute minimum. This sequence, which could have been drawn as a series of two-dimensional diagrams, is far more legible when shown as a number of axonometric slices.

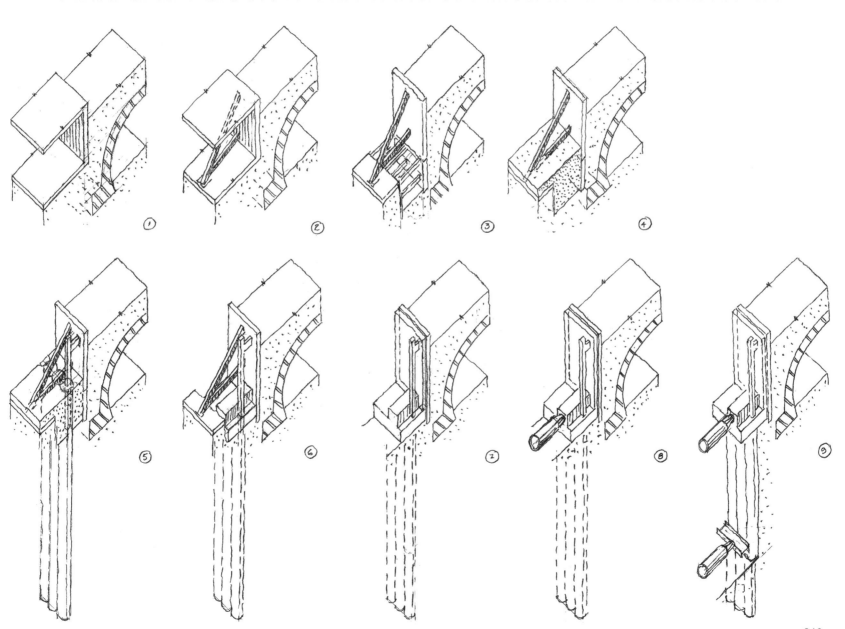

SKETCHBOOK

FIGURE 3.15/4

Kingsgate House, Victoria Street, London. Axonometric drawings show the top-down sequence. Stage 1 is a steel grillage constructed on steel columns plunged into a group of piles. Excavation and the B1 slab are complete (stage 2) followed by construction of the lift pits (stage 3) and then all the intricate detail of the B1, M and ground level slabs complete with base isolation is complete by stage 4.

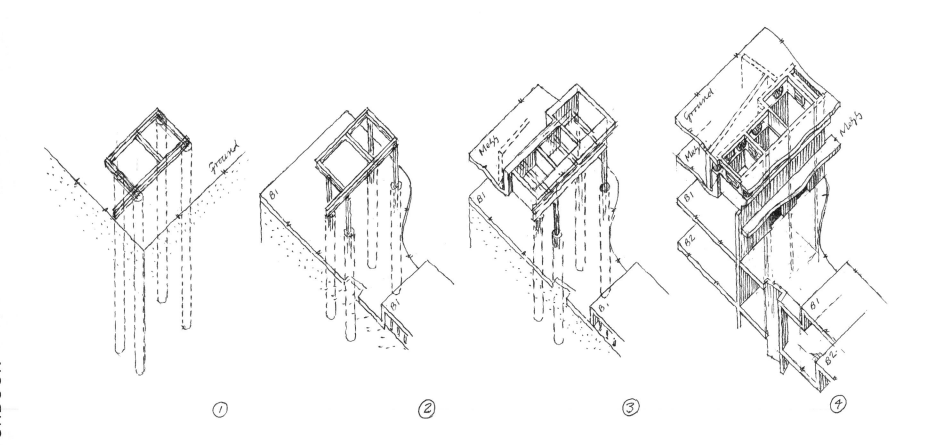

FIGURE 3.15/5

The Shell Centre Redevelopment, London. Built in the late 1950s and early 1960s, the Shell Centre on London's South Bank is scheduled for redevelopment. The big picture depicted here shows the 'U' shaped horseshoe building being demolished but with the tower retained. The basement box or bathtub will also be retained. A new raft is constructed on top of the existing lowest slab and the two structures together will be capable of supporting many of the loads from the new buildings – the need for new piling is therefore minimised.

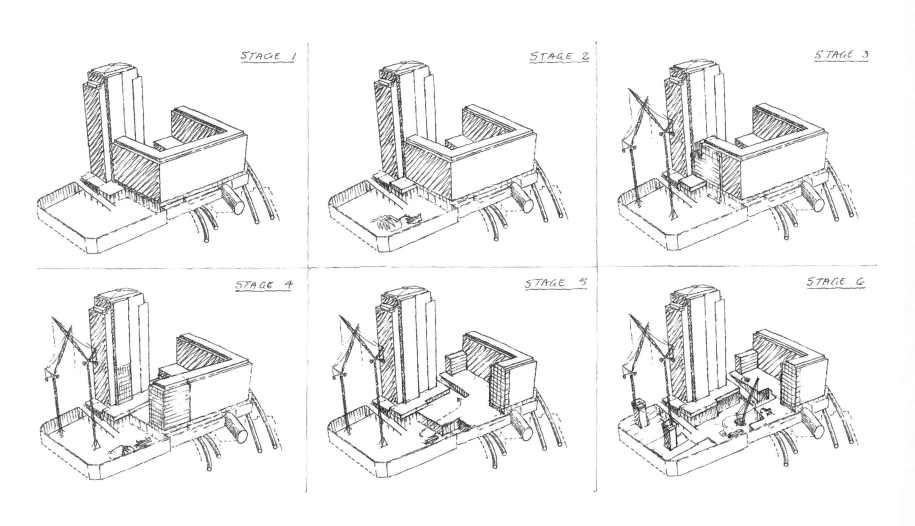

The Place at London Bridge. In some ways, this building by Renzo Piano Building Workshop was far more challenging than its neighbour the Shard. Initial studies quickly showed complexities below ground – running tunnels, escalator declines, vent shafts and escape stairs reduced the available space for foundations to about 50 per cent of the area of a typical upper level floor plate. In the early days, the concept of inclined structure, preset cores designed to resist permanent out-of-balance loads from the building's asymmetry, emerged as the preferred structural solution and in fact was adopted as a key strategy in the final design.

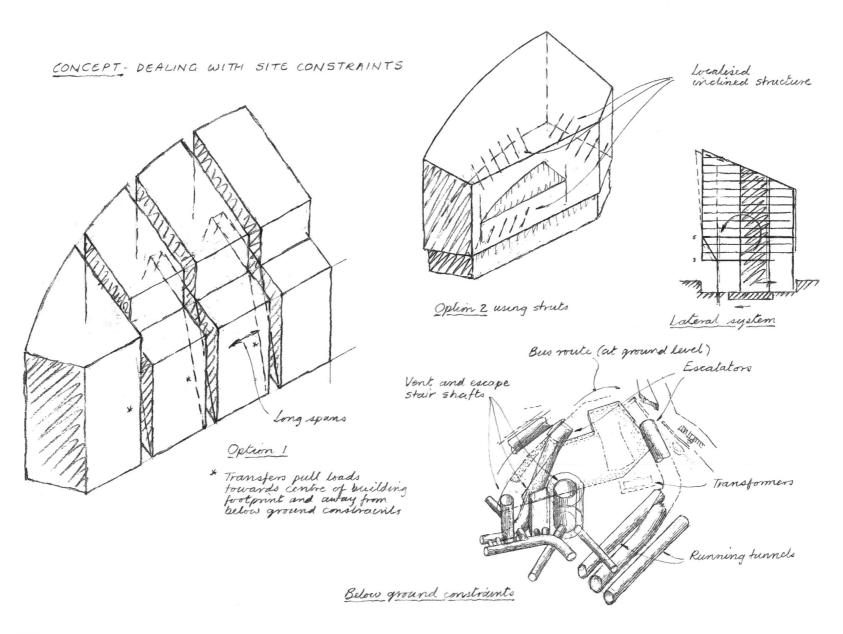

CONCEPT- DEALING WITH SITE CONSTRAINTS

Localised inclined structure

Option 2 using struts

Lateral system

Long spans

Option 1

* Transfers pull loads towards centre of building footprint and away from below ground constraints

Bus route (at ground level)

Escalators

Vent and escape stair shafts

Transformers

Running tunnels

Below ground constraints

FIGURE 3.15/7

Harrods, Knightsbridge, London. Bottom left shows the overview of the project to link Harrods to a newly acquired building east of Basil Street, the former Knightsbridge Crown Court. Knightsbridge Crown Court was redeveloped with a seven-level deep basement behind retained façades and then connected to the main building by lifts and a 5m diameter tunnel. Bottom right shows a grab used to excavate the basement 'top-down'. Top left shows the Crown Court building in more detail and top right is part of many studies to demonstrate buildability on this highly constrained and complex site.

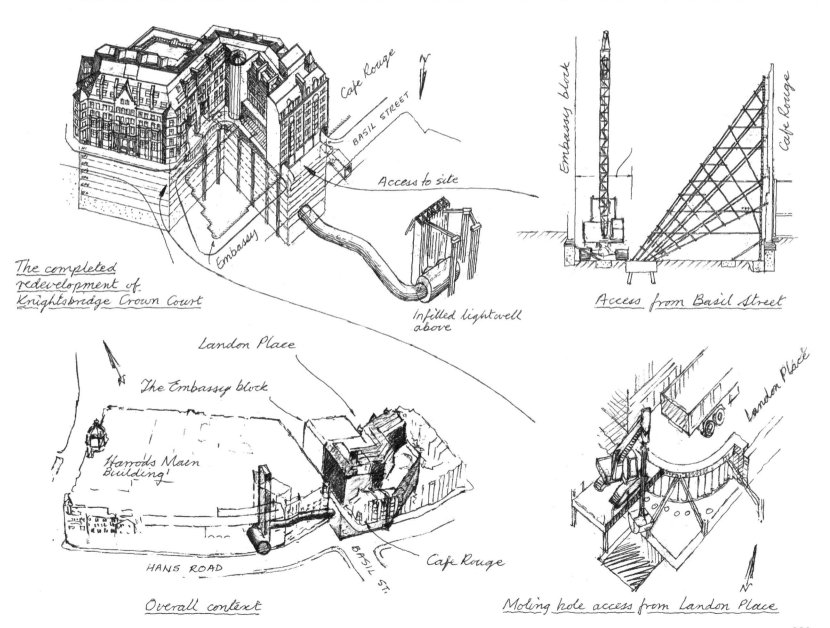

The completed redevelopment of Knightsbridge Crown Court

Infilled lightwell above

Access from Basil Street

Overall context

Moling hole access from Landon Place

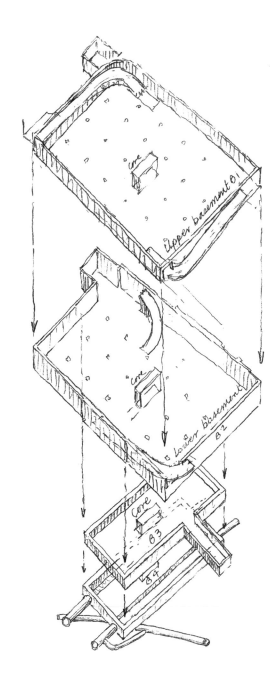

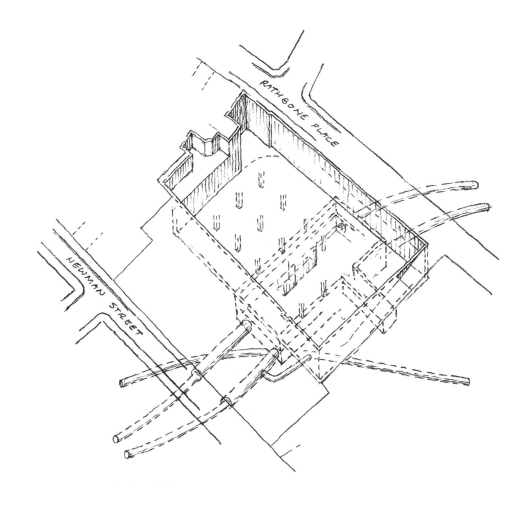

FIGURE 3.15/8

Royal Mail Building, Rathbone Place, London. The drawing on the right shows the basement of the old Royal Mail Building in Rathbone Place, London and the now disused Royal Mail Mail Rail station and tunnels. New development must preserve the station and the tunnels but the upper basement itself is to be reconfigured. Spaced apart layers shown on the left, is the clearest way of providing an instant picture of the new proposals.

FIGURE 3.15/9

199 Knightsbridge, London. Working around flying shores is difficult but on a relatively narrow site where new construction is founded on a raft, there is little alternative. Omitting the flying shores by using cantilevered embedded perimeter walls would be uneconomic and top-down construction for a two- or three-level basement, which requires a piled foundation instead of a raft, would also not be economic. This sketch was part of a construction sequence.

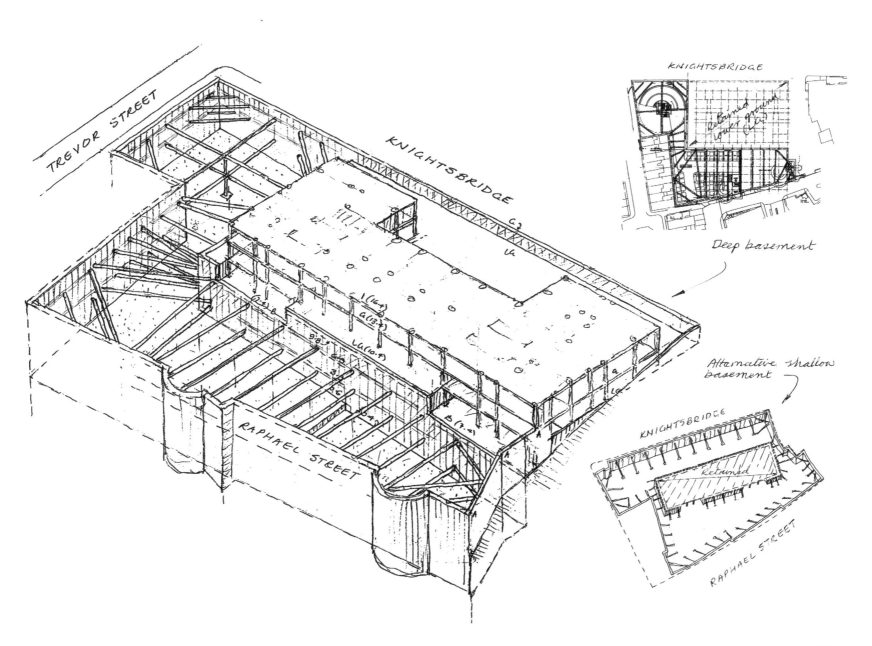

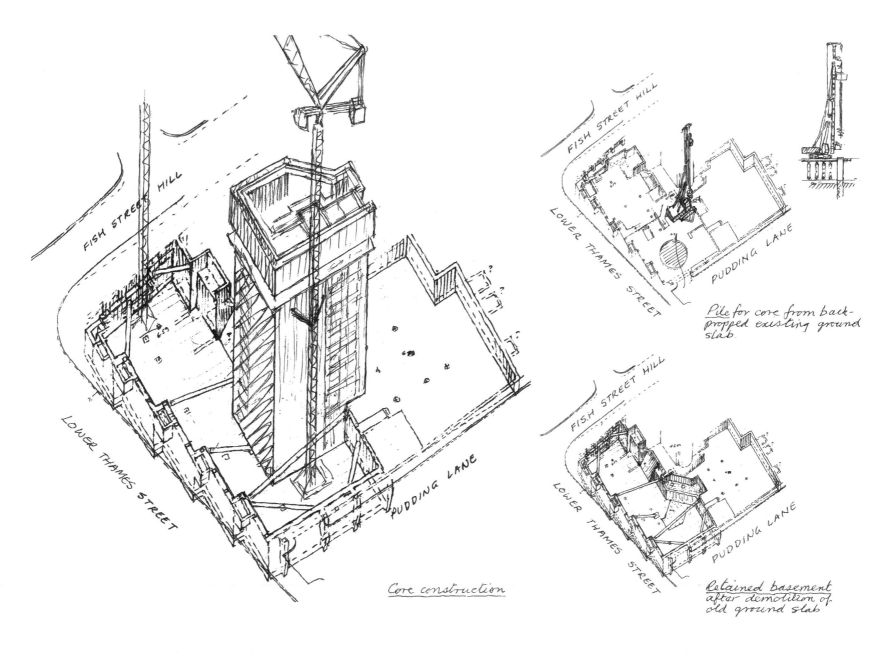

Pile for core from back-propped existing ground slab.

Core construction

Retained basement after demolition of old ground slab

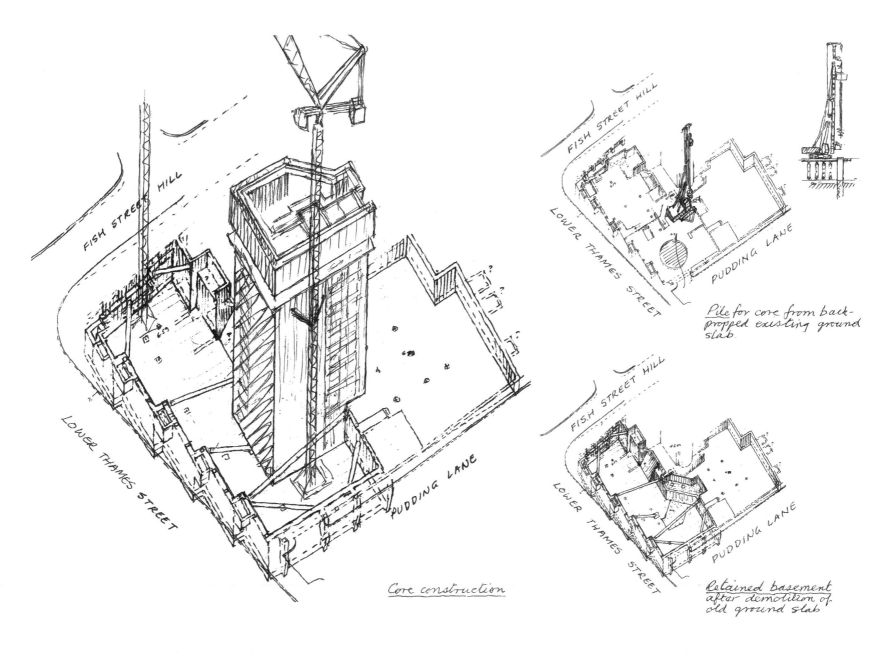

FISH STREET HILL

LOWER THAMES STREET

PUDDING LANE

FIGURE 3.15/10

Centurion House, Lower Thames Street, London. Monument Place is built on the site of the northern abutment of the first Roman crossing of the Thames, details of which were revealed by archaeological excavations carried out prior to the construction of a 1960s building. It was important that redevelopment in 2010–13 caused as little disturbance as possible to the archaeology and consequently only the core is founded on new piles. The rest of the building is supported on reused substructure and piles from the 1960s building.

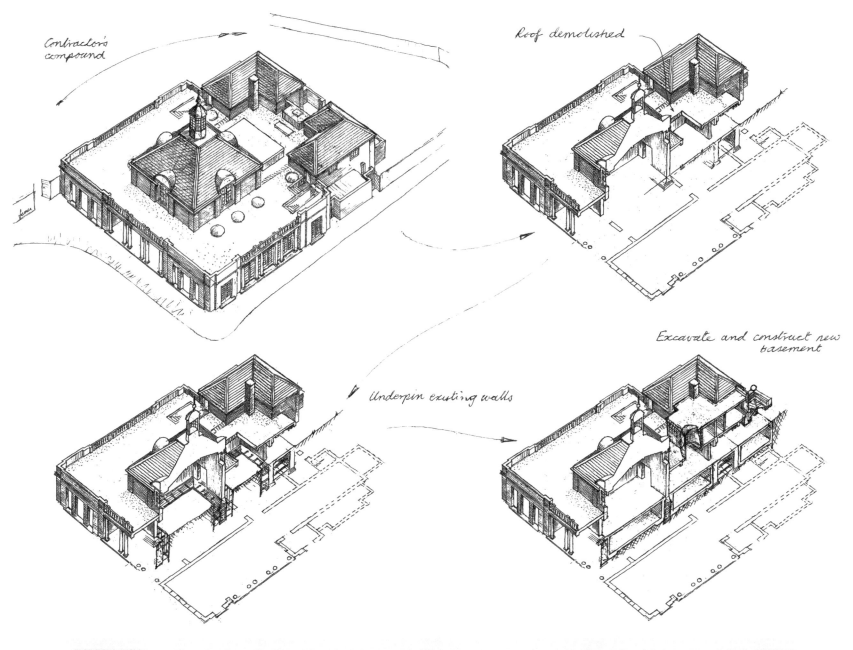

Contractor's compound

Roof demolished

Underpin existing walls

Excavate and construct new basement

FIGURE 3.15/11

The Serpentine Gallery, Kensington Gardens, London. Basement construction in urban areas is always challenging, especially below an existing building and even more so under a listed nationally important art gallery. Symmetry of the existing building and of the new works meant that cutaway sequence drawings were easily produced and provided a reasonably complete picture of the proposed methodology.

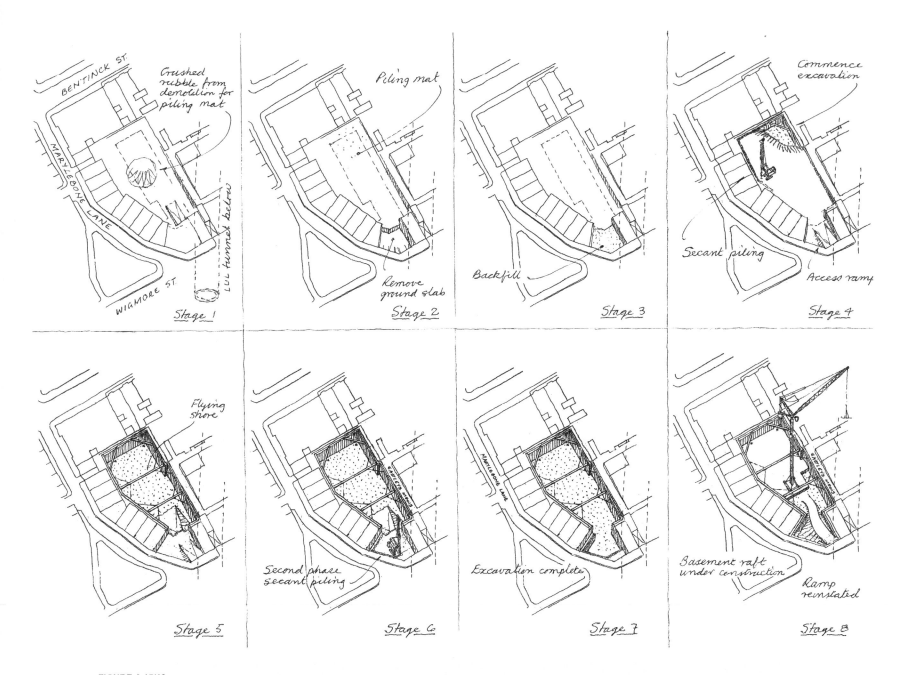

FIGURE 3.15/12

Wigmore Street, London. For the sake of clarity it is sometimes best just to show only the ground level footprints of adjacent buildings, especially on landlocked sites such as this development on Wigmore Street. Logistics, site access, party wall issues and the general effect on neighbours are then more easily investigated. Separate sketches can be developed quickly showing surrounding basements and underground infrastructure.

Riverwalk, London. Riverwalk is a high-end residential development at Millbank near Vauxhall Bridge. It consists of two curvilinear buildings constructed over a common basement. Complex structural slabs deal with the transfer of loads from columns positioned in the superstructures to suit the layout of the apartments, to acceptable positions in the below-ground car park. There are in fact two transfer slabs, one at ground level and another over the car ramp.

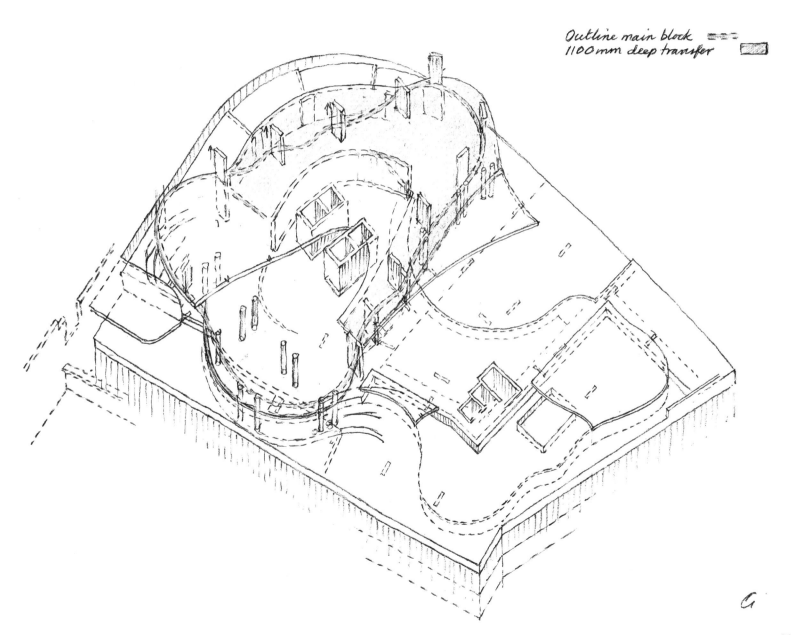

Outline main block
1100 mm deep transfer

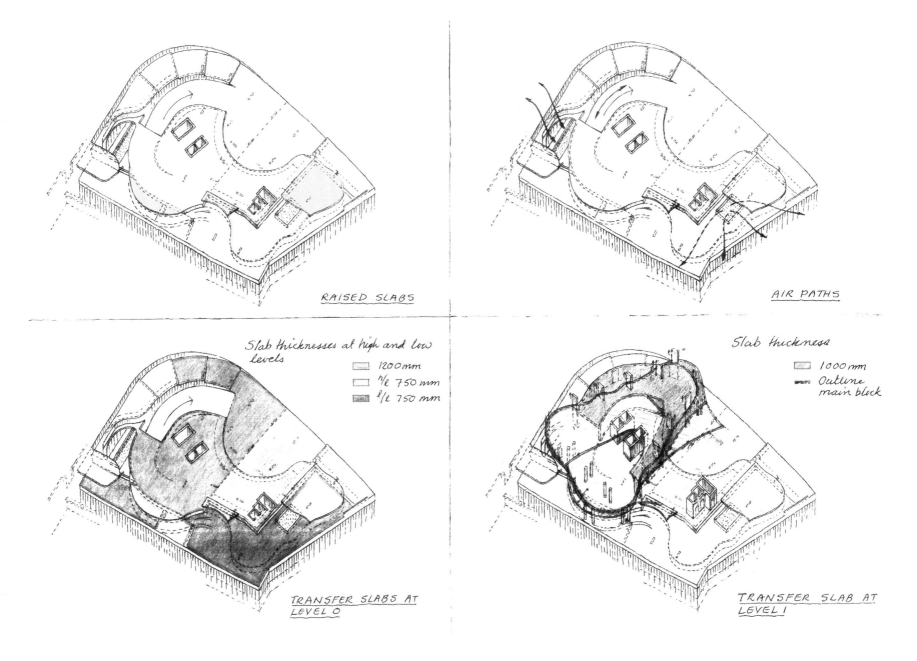

RAISED SLABS

AIR PATHS

Slab thicknesses at high and low levels

▭ 1200 mm
▭ ʰ/ₑ 750 mm
▭ ˡ/ₑ 750 mm

Slab thickness

▭ 1000 mm
▰ Outline main block

TRANSFER SLABS AT LEVEL 0

TRANSFER SLAB AT LEVEL 1

FIGURE 3.15/14

Riverwalk, London. To understand the design principles behind the Riverwalk slabs it was necessary to first draw the lower transfer (bottom left) using different weight shading to show various structural thicknesses, and then draw the level 1 car park ramp transfer separately (bottom right). Details could then be explained, for instance the addition of a raised slab to form a balcony and a raised slab to form an air plenum (shaded, top left) – the same diagram was used to show non-structural functionality (top right).

Riverwalk, London. Riverwalk is right beside the Thames and the lowest parts of the basement are below high tide level. Groundwater is hydraulically linked to the river and therefore the design of a high quality perimeter secant wall was essential to keep the excavation dry and safe. Controlling deflection of the wall using knee braces and flying shores was part of the strategy and is represented here using a straightforward axonometric sketch.

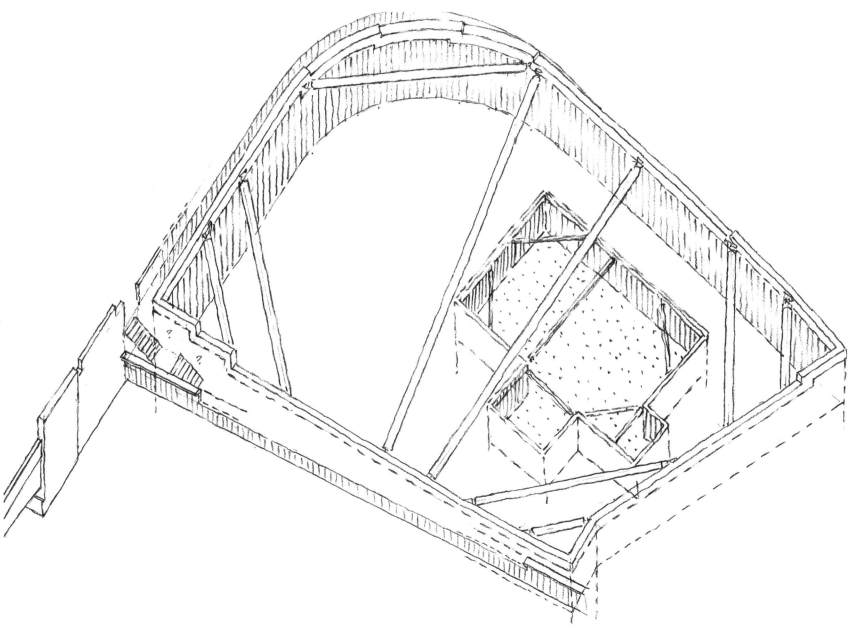

FIGURE 3.15/16

Bulgari Hotel, Knightsbridge, London. The Bulgari Hotel was the first development in the UK to use a combination of piles and basement walls to provide ground source heating and cooling. Water is circulated through a system of embedded pipework and the proximity of the pipework to the sub-soils, which are at a near constant temperature throughout the year, either cools the circulating water in summer or warms it in winter.

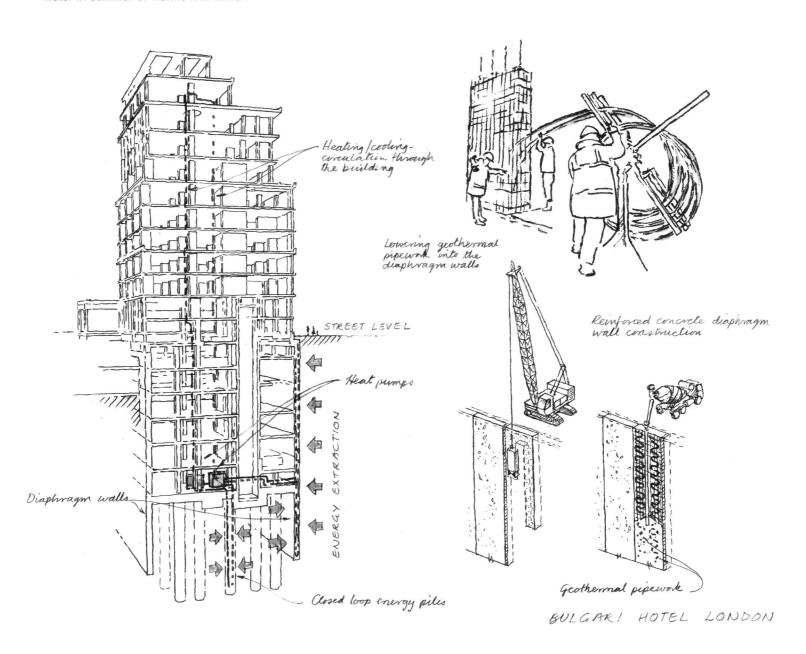

Heating/cooling-circulation through the building

Lowering geothermal pipework into the diaphragm walls

STREET LEVEL

Heat pumps

ENERGY EXTRACTION

Diaphragm walls

Reinforced concrete diaphragm wall construction

Closed loop energy piles

Geothermal pipework

BULGARI HOTEL LONDON

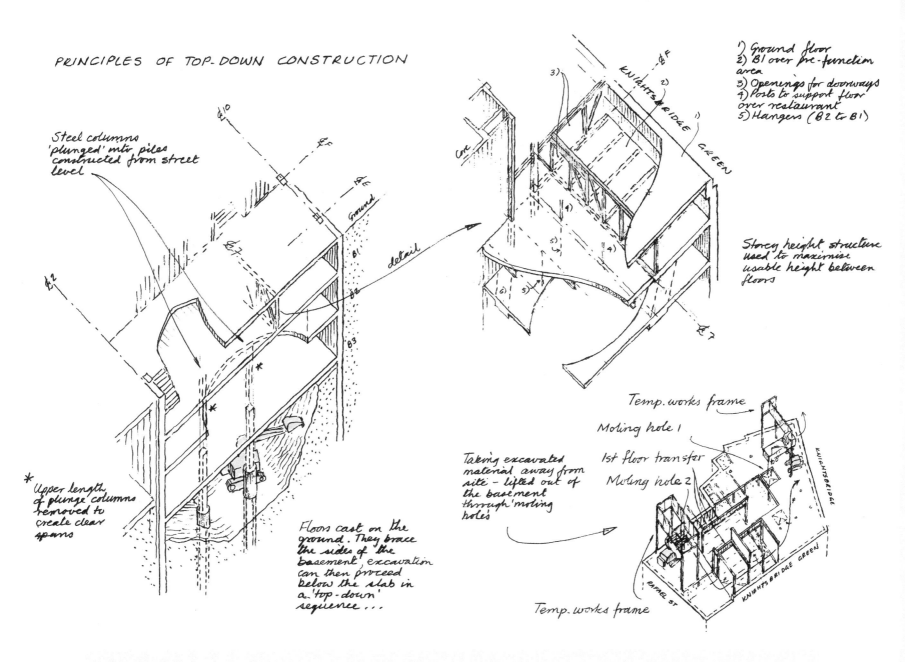

FIGURE 3.15/17

Bulgari Hotel, Knightsbridge, London. The principles of top-down construction are best explained using sketches, especially if the sequence is unusual. Normally steel columns are 'plunged' into newly concreted piles, then the ground floor is cast, it props the perimeter basement walls, excavation is carried out below the floor and the sequence is repeated level by level for the lower floors. At the Bulgari, large column-free areas were required so transfer structure had to be built into some of the floor structures, allowing a number of the top-down columns to be cut out after performing their temporary function.

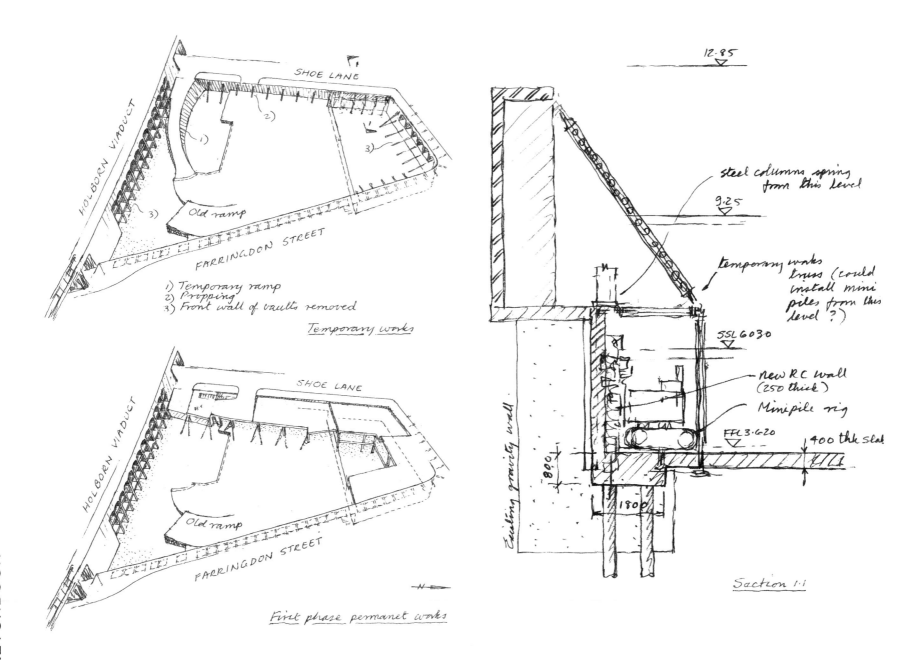

FIGURE 3.15/18

Atlantic House, Holborn, London. Work in the ground at Atlantic House proved to be incredibly difficult due to the presence of heavy foundations from earlier buildings. The Shoe Lane and Charterhouse perimeters had to be propped during foundation and substructure construction, whereas the Holborn Viaduct and Farringdon Street perimeters were largely self-supporting. Another factor that slowed the project significantly was the discovery of significant Roman remains.

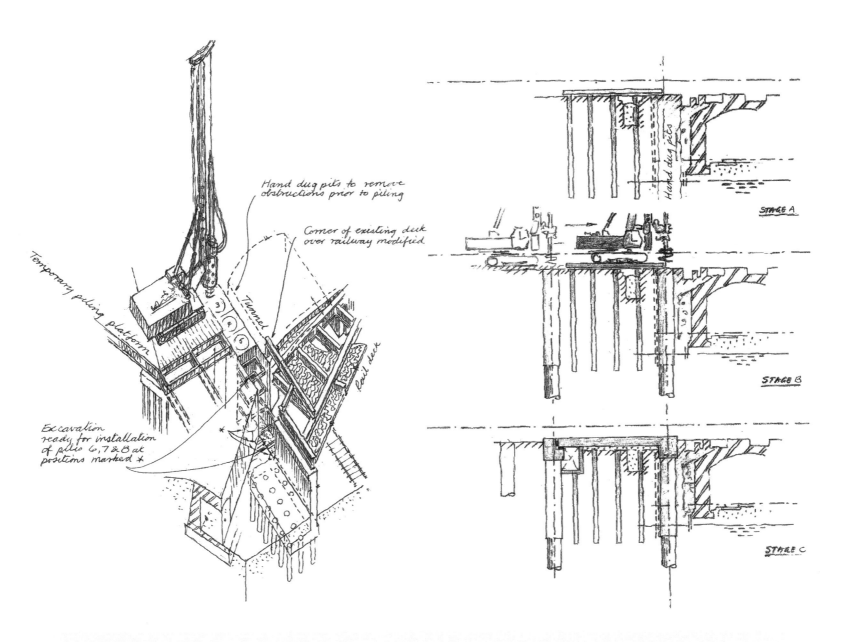

Hand dug pits to remove obstructions prior to piling

Corner of existing deck over railway modified

Temporary piling platform

Tunnel

Rail deck

Excavation ready for installation of piles 6, 7 & 8 at positions marked ✱

STAGE A

STAGE B

STAGE C

Hand dug pits

FIGURE 3.15/19

Principal Place. Large diameter augered piles were constructed close to the existing tunnel as part of the foundation for the residential tower at Principal Place, London. Their precise position was dictated by the existing tunnel and an adjacent corridor that is allocated for future railway lines. The methodology for getting these piles in the ground involved hand dug pits to remove obstructions, construction of a piling platform for the 180-tonne rig, itself supported on small diameter piles, and a very specific sequence for digging the pits and constructing the piles. A series of axonometric sketches was produced together with two-dimensional diagrams to interrogate the spatial fit and to illustrate the basic concept.

FIGURE 3.15/20

Principal Place. The sketch on the left was produced to show a proposal for tackling a particularly tricky corner of the site at Principal Place in London, but standing a piling rig on a road near a railway is always difficult to organise and agree with the rail and highways authorities. A completely different strategy was devised with the contractor to retain the road using a gravity block instead of embedded contiguous piles. The 'plan B' sketches on the right were developed in the site office using tracing paper over a single photograph – far easier and quicker than using CAD.

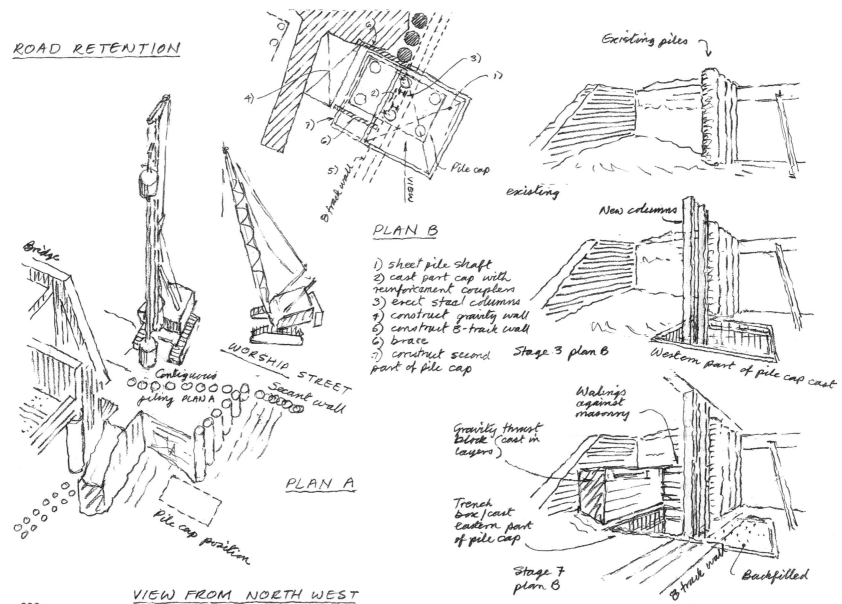

ROAD RETENTION

PLAN B

1) sheet pile shaft
2) cast part cap with reinforcement couplers
3) erect steel columns
4) construct gravity wall
5) construct 8-track wall
6) brace
7) construct second part of pile cap

PLAN A

VIEW FROM NORTH WEST

Existing piles

existing

New columns

Stage 3 plan B

Western part of pile cap cast

Walings against masonry

Gravity thrust block (cast in layers)

Trench box / cast eastern part of pile cap

Stage 7 plan B

8-track wall

Backfilled

Bridge

WORSHIP STREET

Contiguous piling PLAN A

Secant wall

Pile cap position

FIGURE 3.15/21

Royal Scottish Academy, Edinburgh, Scotland. Shallow tunnelling is a difficult process but has to be considered when open cut techniques are precluded by the presence of vital services and the need to keep traffic flowing. Conventional two-dimensional sections and axonometric sketches were used to show the engineering involved in Terry Farrell's scheme to link Edinburgh's western gardens to the National Gallery of Scotland and the Royal Scottish Academy.

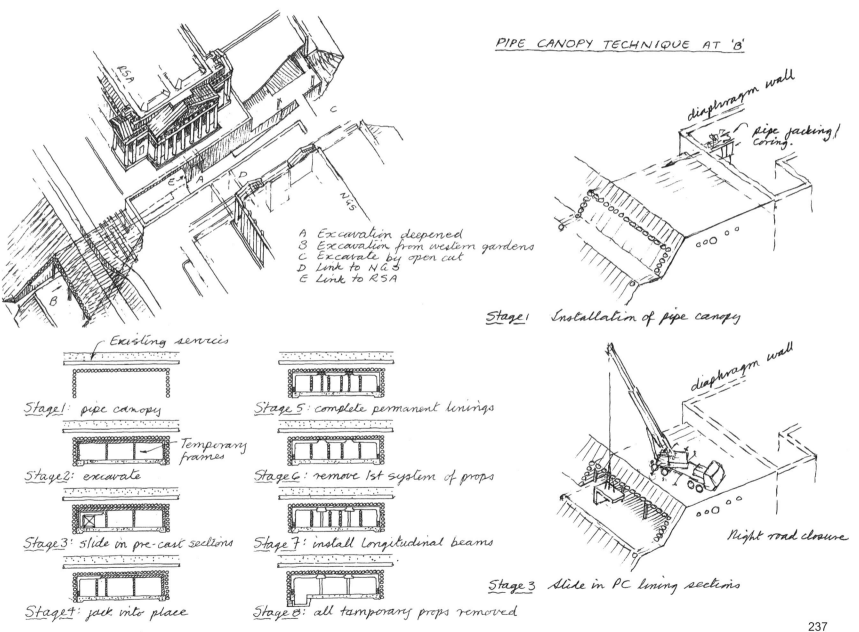

PIPE CANOPY TECHNIQUE AT 'B'

diaphragm wall

pipe jacking / coring.

A Excavation deepened
B Excavation from western gardens
C Excavate by open cut
D Link to N G S
E Link to RSA

Stage 1   Installation of pipe canopy

diaphragm wall

Night road closure

Stage 3 Slide in PC lining sections

Existing services

Stage 1: pipe canopy

Temporary frames

Stage 2: excavate

Stage 3: slide in pre-cast sections

Stage 4: jack into place

Stage 5: complete permanent linings

Stage 6: remove 1st system of props

Stage 7: install longitudinal beams

Stage 8: all temporary props removed

237

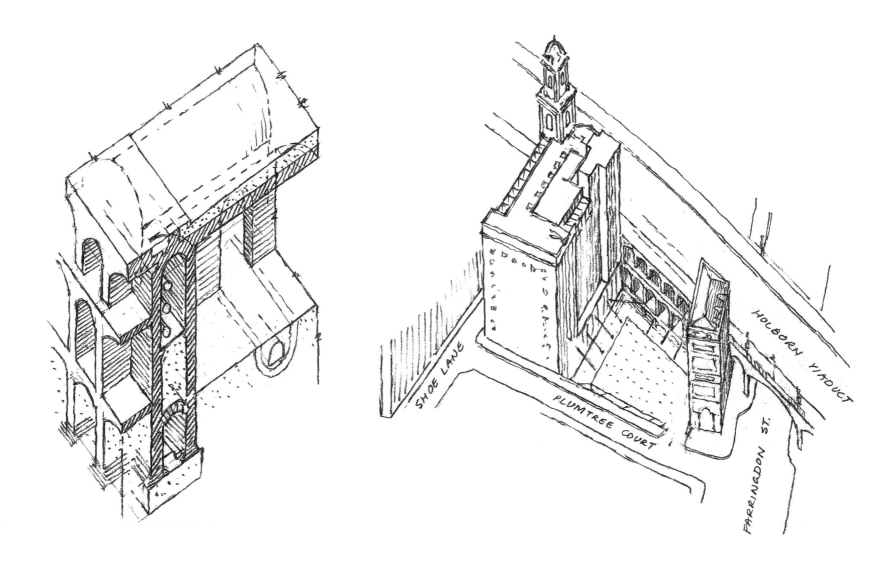

FIGURE 3.15/22

Morley House, London. Holborn Viaduct in London really is a viaduct and not just a bridge over Farringdon Street – the approach roads are constructed on multi-level arches. It means the road beside this site is self-supporting and doesn't require extensive propping or shoring. A portion of the viaduct makes a satisfying piece of three-dimensional geometry and drawing it is almost irresistible.

FIGURE 3.15/23

Admiralty Arch, London. Admiralty Arch may be a very familiar building but it has some surprises in store – first, the shape, difficult to read from the well-known Trafalgar Square or Mall picture-postcard views, and second the existence of complicated below-ground infrastructure. The old Fleet Line tunnels, still in use as sidings for training purposes and occasionally for filming, are quite shallow and are directly below the road. In fact, right here, they are joined together by a large diameter cross-over tunnel. Prime Investors Capital Ltd is planning to turn the disused building into a luxury hotel by constructing new basements fore and aft of the Arch.

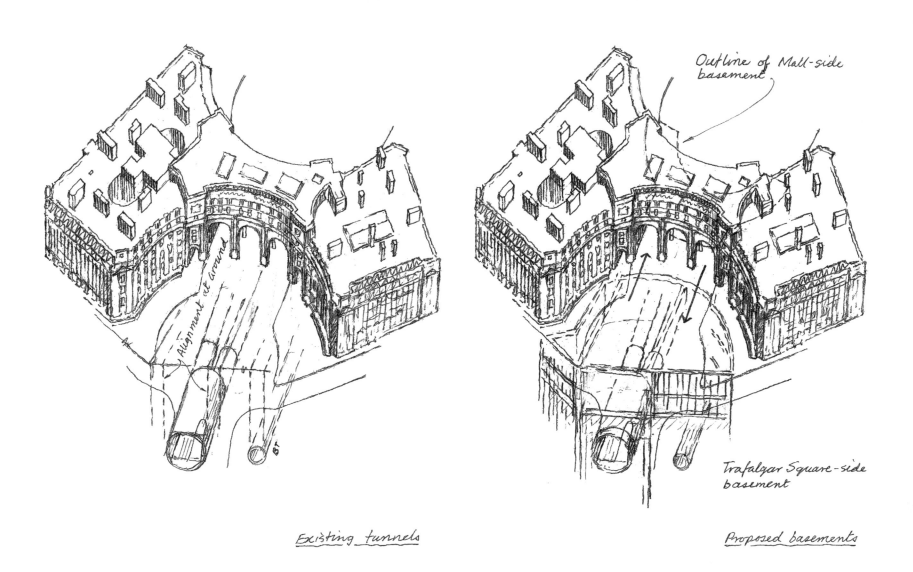

Existing tunnels

Outline of Mall-side basement

Trafalgar Square-side basement

Proposed basements

3.16 SUSTAINABLE STRUCTURES

Sustainable design of buildings can have a significant effect on construction and running costs as well as having a positive environmental impact. In-use energy costs can be reduced by up to 75 per cent. Sustainability is now part of the normal design process and sometimes has a major influence on the architectural and structural form of a building. When it does, architects and engineers develop concepts in close collaboration by exchanging ideas. Sketching becomes an essential and rewarding part of the design process. Whether design is driven by sustainability principles or not, it is worth remembering the basic strategies that engineers can employ:

Construction materials

- Cement replacement (ggbs and/or pfa)
- Recycled concrete as aggregate
- High strength concrete
- Self-compacting concrete
- Reinforcement couplers
- Pre-stressed concrete
- High strength steel
- Responsibly resourced materials.

Groundwork

- Minimise spoil taken off site
- On-site decontamination
- Avoid piling
- Re-use existing foundations
- Displacement auger piling
- Energy piles: cooling/heating fluid circulates through pipes cast in piles

- Geothermal heat pumps
- Sustainable drainage systems (SuDS).

Superstructure

- Retain existing structures
- Optimise floor loadings
- Relax deflection and vibration limits
- Lean construction
- Composite columns (concrete-cased sections, concrete-filled tubes)
- Fabric thermal storage
- Thermally broken balconies
- Life cycle costing.

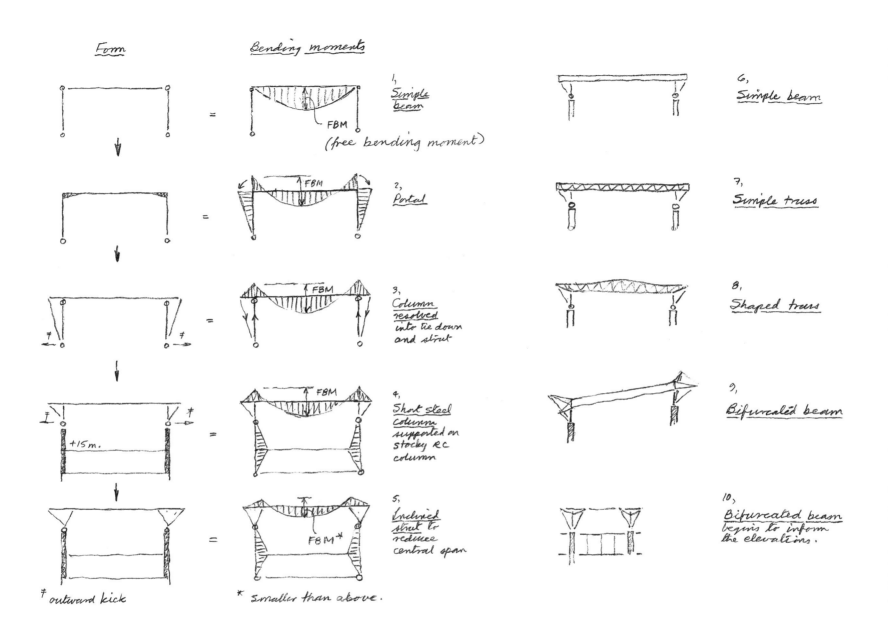

Form   **Bending moments**

1, Simple beam
FBM (free bending moment)

2, Portal
FBM

3, Column resolved into tie down and strut
FBM

4, Short steel column supported on stocky RC column
FBM
+15m.

5, Inclined strut to reduce central span
FBM *

† outward kick

* smaller than above.

6, Simple beam

7, Simple truss

8, Shaped truss

9, Bifurcated beam

10, Bifurcated beam begins to inform the elevations.

FIGURE 3.16/1

Biota! Silvertown Quays Aquarium. Solutions for the 'biome' roofs ranged from simplistic 'beam and post' arrangements stabilised by a combination of roof bracing and perimeter shear frames, to structures based on portal frame variants. The frames were used to span the width of the building while stability shear frames were built into the long elevations. Analyses of the different portal frame options were shared with the architect in a search for aesthetically pleasing and efficient structure. The horizontal members of the portals developed into two-dimensional and then three-dimensional trusses which might in turn, have influenced the design of the elevations.

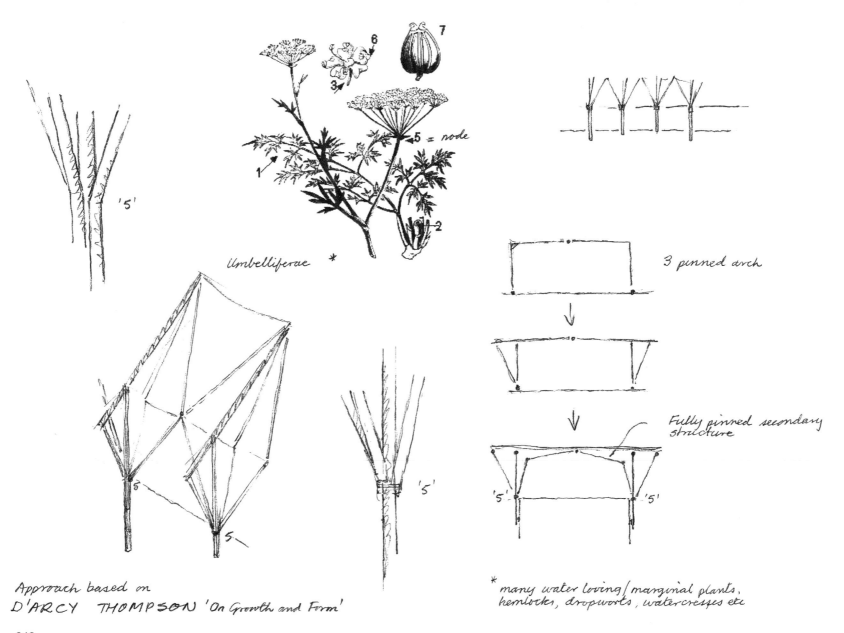

FIGURE 3.16/2

Biota! Silvertown Quays Aquarium. The general massing of the building was largely dictated by the shape of the site, and by the exhibit designers whose overriding objective was to create a logical, informative and entertaining route for visitors through the displays. Plan shape was chosen to tie in with local landscape features including the Thames Barrier Park but apart from this, much of the conventional architectural design was focussed on the naturally lit upper levels of the building, the 'biome' roofs. Many options were explored for the structures, some tree-like and others based on more ambitious biological forms.

FIGURE 3.16/3

Biota! Silvertown Quays Aquarium. One of the more interesting proposals for the roof trusses for the biomes consisted of a regular Warren truss constructed using a combination of materials – Glulam for the compression members and steel for the tension members. Detailing of visual connections was partially developed and geometry and setting out of the truss internal members selected to ensure the more heavily loaded compression members near the supports were the shortest and best able to cope with local buckling. The design was predicated on there being sufficient dead load to counteract wind uplift loadings.

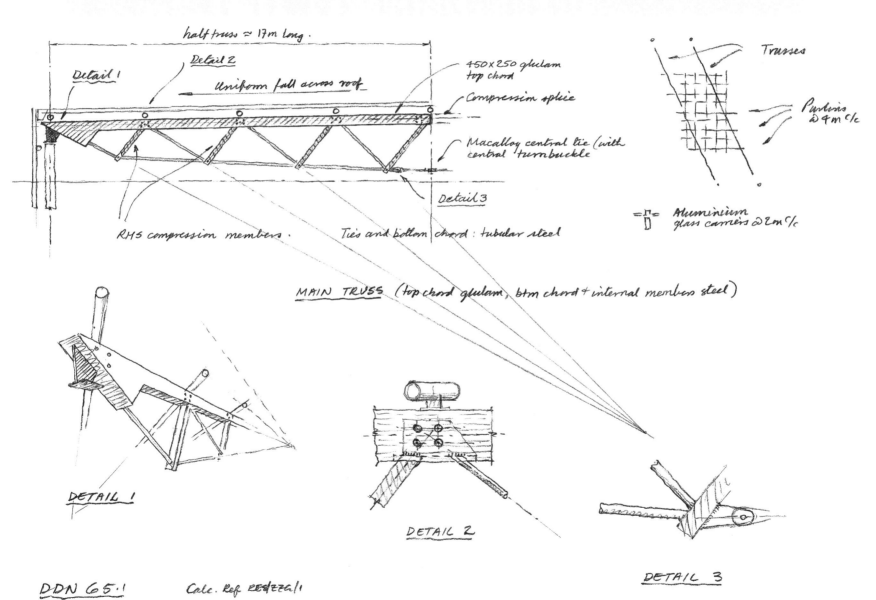

half truss ≈ 17m long.

Detail 1

Detail 2

Uniform fall across roof

450 × 250 glulam top chord

Compression splice

Macalloy central tie (with central turnbuckle)

Detail 3

RHS compression members.

Ties and bottom chord : tubular steel

MAIN TRUSS (top chord glulam, btm chord + internal members steel)

Trusses

Purlins @ 4m c/c

Aluminium glass carriers @ 2m c/c

DETAIL 1

DETAIL 2

DETAIL 3

DDN 65·1   Calc. Ref REDZZA/1

WSPCS   8/3/04

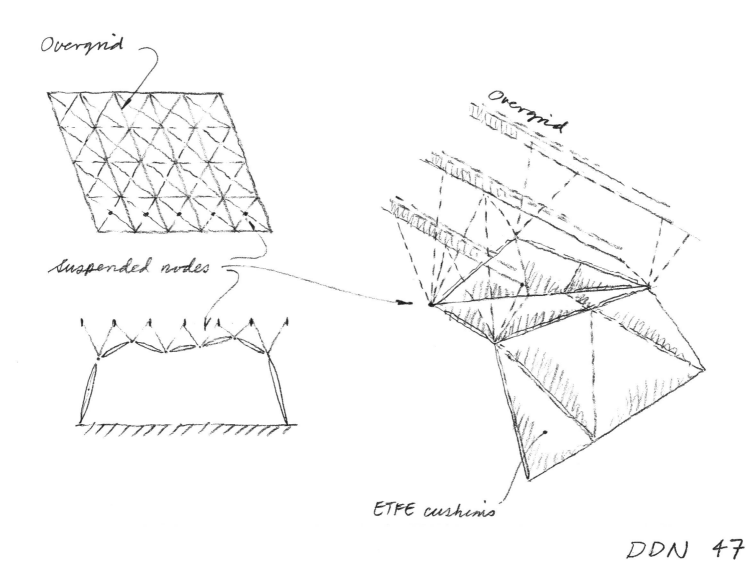

BIOME CONCEPT

Overgrid

Overgrid

suspended nodes

ETFE cushions

DDN 47

SKETCHBOOK

FIGURE 3.16/4
Biota! Silvertown Quays Aquarium. The final proposal for the biome roofs owed little to earlier design studies; if anything the design had reverted to the very first 'beam and post' concept. The over-grid of beams supported a chaotic arrangement of triangular ETFE cushions. Nodes of this super-lightweight envelope were to be held in position by thin tubular members designed for snow and wind uplift loads.

244

FIGURE 3.16/5

Shallow foundation option 1 of 2, Middle East. This foundation was for a Middle Eastern sustainable, mixed-use development designed to be friendly to pedestrians, cyclists and the environment in general. It has terracotta walls decorated with arabesque patterns modelled on ancient cities and is designed with short, narrow streets arranged to create a cooling, flushing effect. Foundation options were studied to find the simplest, most efficient solution based on stabilising existing soils rather than piling to rock-head.

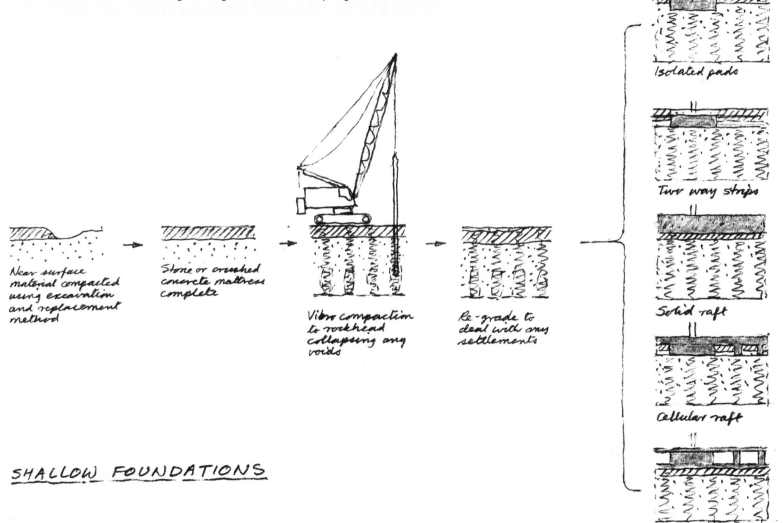

OPTIONS

Isolated pads

Two way strips

Solid raft

Cellular raft

Inverted raft.

Near surface material compacted using excavation and replacement method

Stone or crushed concrete mattress complete

Vibro compaction to rockhead collapsing any voids

Re-grade to deal with any settlements

SHALLOW FOUNDATIONS

Shallow foundation option 2 of 2, Middle East. A cellular raft option was developed as a robust but flexible structure capable of supporting heavy on-grid column loads and lighter off-grid loads. It was engineered to spread loads to improved natural ground and to accommodate a network of service trenches. The aim was to construct a foundation that would accommodate future change and minimise cost, volume of concrete and embodied carbon.

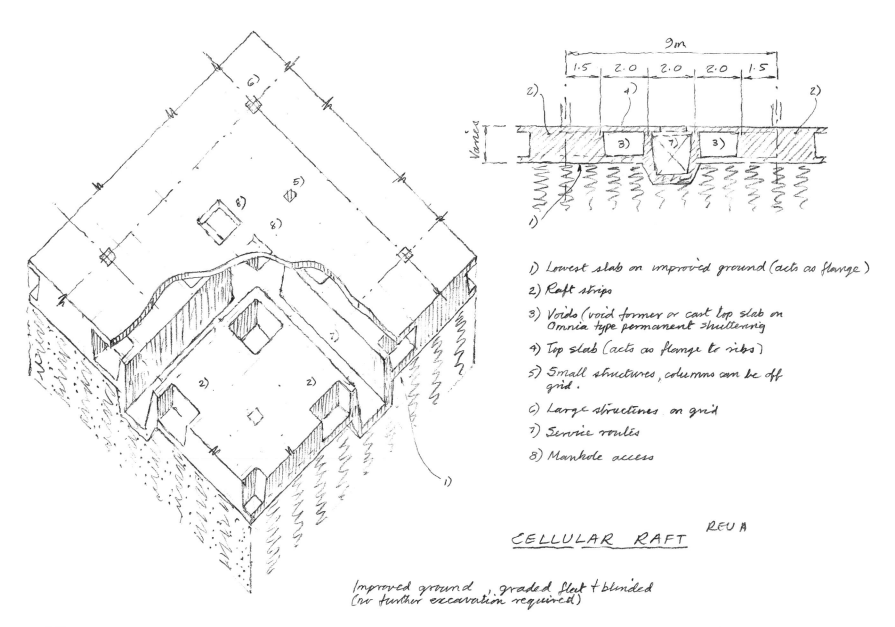

1) Lowest slab on improved ground (acts as flange)
2) Raft strips
3) Voids (void former or cast top slab on Omnia type permanent shuttering)
4) Top slab (acts as flange to ribs)
5) Small structures, columns can be off grid.
6) Large structures on grid
7) Service routes
8) Manhole access

CELLULAR RAFT   REV A

Improved ground, graded flat + blinded (no further excavation required)

Penarth. Ideas for a green seaside resort at Penarth, South Wales – just a rough
sketch trying to capture a number of ideas.

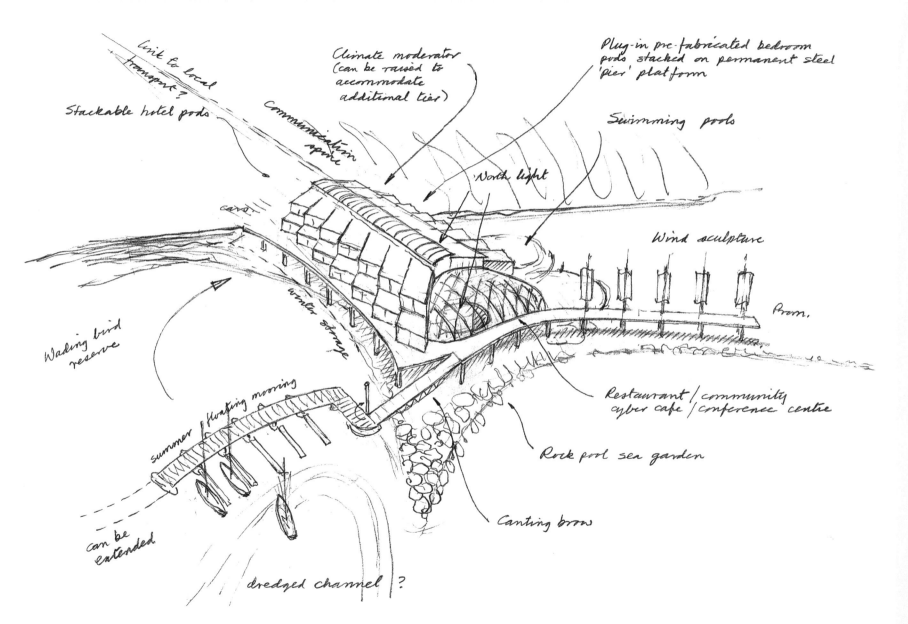

Link to local transport?

Stackable hotel pods

Climate moderator
(can be raised to
accommodate
additional tiers)

Communication space

North light

Plug-in pre-fabricated bedroom pods stacked on permanent steel 'pier' platform

Swimming pools

cars

Wind sculpture

Winter storage

Prom.

Wading bird reserve

summer floating mooring

Restaurant / community cyber cafe / conference centre

Rock pool sea garden

can be extended

Canting brow

dredged channel ?

Bulgari Hotel, Knightsbridge, London. Structural engineers don't often get an excuse to draw wildlife but green and brown roofs have an impact on loadings and it's good to show we are at least aware of other things like the need for bat boxes and swift nesting boxes.

GREEN ROOFS....

Carbon-neutral warehouse. A major logistics company discovered that they could not build in Continental Europe without proposing credible carbon-neutral designs for their major distribution centres. The typical bay shown here was a response to those difficulties based on naturally lit interiors (through ETFE cushions), green roofs, pre-cast hyperclastic concrete roof units, pre-cast dock leveller units and cladding, and 'sensible spans' to provide an efficient and lean structure. This was all backed up by plans that included responsible sourcing of materials and free draining roads and hardstandings.

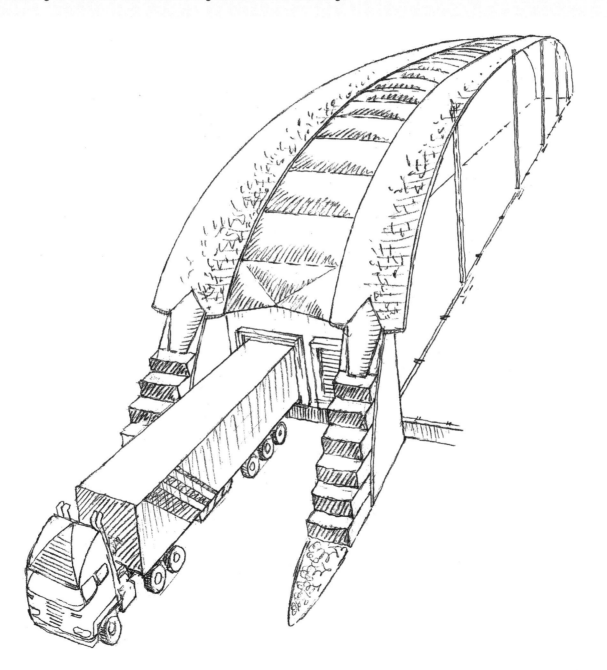

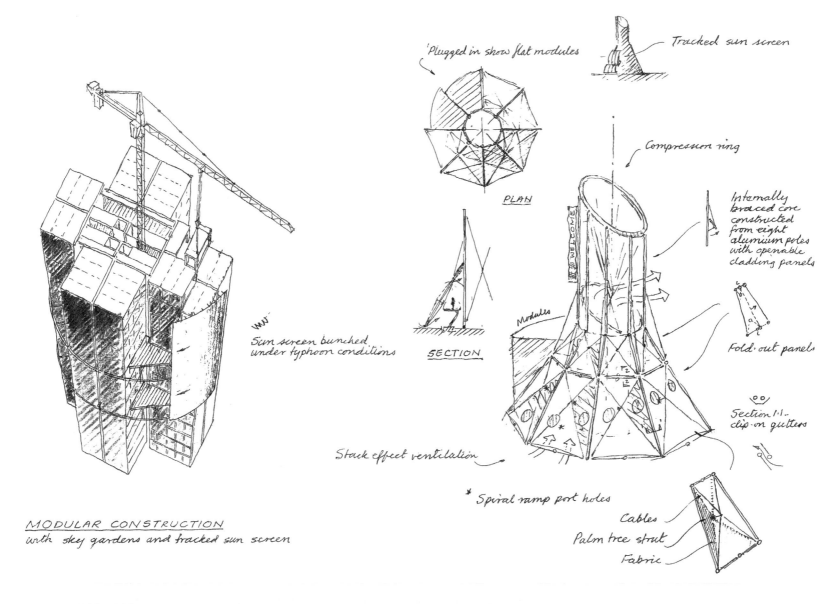

'Plugged in show flat modules

Tracked sun screen

Compression ring

PLAN

Internally braced core constructed from eight aluminium poles with openable cladding panels

Sun screen bunched under typhoon conditions

SECTION

Modules

Fold·out panels

Section 1·1· clip·on gutters

Stack effect ventilation

* Spiral ramp port holes

Cables
Palm tree strut
Fabric

MODULAR CONSTRUCTION
with sky gardens and tracked sun screen

FIGURE 3.16/10

Temporary visitor centre, Hong Kong. This pavilion structure was designed as part of a project to promote innovation in residential development in Hong Kong. The pavilion exhibited two apartment 'building blocks' or modules intended for use in the sustainable construction of multi-storey residential towers. The INTEGER organisation was commissioned to design an exhibition which would not only demonstrate sustainability but would also explain how new ideas could be used to raise the quality of the built environment. Sustainability principles were adopted to design and construct the pavilion, including off-site prefabrication, the use of recycled materials, stack effect and natural ventilation. The pavilion was built on the waterfront in Hong Kong at Tamar, and true to the aspirations of the design team, was dismantled and reused on a site in Beijing.

PART 4

EPILOGUE

It is clear that sketching and drawing have played a major part in socio-economic and scientific development from the earliest times, probably even before the days when we protected ourselves from wild animals, our enemies and the weather in dark and cold caves. Over the few thousand years since those primitive days, we have learnt new skills and invented new techniques but the value of transferring an image from our heads to a physical surface or developing an idea by reworking it in front of our eyes, is still immense.

Spencer Frederick Gore (1878–1914) sums it up as follows: '*By drawing, man has extended his ability to see and comprehend what he sees*'.

And there's the famous 'A picture is worth a thousand words' quotation. This phrase emerged in the USA in the early part of the twentieth century. Its introduction is widely attributed to Frederick R. Barnard, who published a piece commending the effectiveness of graphics in advertising with the title 'One Look Is Worth a Thousand Words', in *Printer's Ink* of December 1921. Barnard claimed the source of the phrase to be oriental by adding the text 'so said a famous Japanese philosopher, and he was right'.

In recent years, it has been a disappointment to walk around the design office, once known as the drawing office, and find it difficult to spot a pencil, let alone a drawing board. But all is not lost, people like to draw and sketch and are easily convinced that it is still an essential communication skill which will help them to get more out of their day to day lives. It is no accident that sketching and drawing survives in many walks of life, but it is worth reminding ourselves, in an age of computers, that the simplest way of sketching, using pencil and paper, has its place in our frenetic engineering and architectural world.

Happily throughout my career, some things have not changed: despite all the frustrations involved in making things happen, there has always been great satisfaction gained from getting worthwhile projects built. An essential part of the process of building things is to communicate with others: no one person designs or builds a building, we are always part of a team. Being part of a team is the best and most motivating experience of all, and communicating with team members is therefore vitally important. Sketching is often the best way of communicating.

Drawing and sketching is a means of developing concepts and thinking through strategies and details. Problems are solved en route and construction is understood. It is an iterative process that takes time, persistence and patience but the end result is worth the effort. As an important bonus, producing a sketch can in itself be treated as a project, and with the right amount of diligence, the end result will be highly satisfying even though it may not be a work of art.

We can all improve our skills and abilities with a little bit of effort and perseverance. For design related to the built environment there can be few better ways to communicate than by sketching, and we are all capable of sketching, whether we have artistic talent or not.

In some ways Part 3 is a record of my engineering career starting in the 1960s through to the early years of the twenty-first century, from the time when engineers and architects were beginning to use calculators to an era where almost everyone uses sophisticated computer magic, and I hope it demonstrates that hand sketching has played a big part in it. It also reflects my career, the last fifteen years of which have been focused on high rise, whereas for me, projects in other sectors such as marine developments have been very few and far between.

Admittedly there is more to developing a successful career in engineering or architecture than sketching – but how do we judge a successful career? It must be about achieving goals, by working with others in a team; it is about the practical realisation of ideas, concepts and dreams. Sketching will play its part.

Tony Hunt has put it more eloquently:

> As a young engineer I discovered that describing a structural concept or detail wasn't very effective unless illustrated with a sketch to explain my ideas. My mind has the ideas, but they only become 'real' through the medium of sketching, since mental ideas, as yet, cannot be transmitted to others. And anyway, the act of drawing what the mind 'sees' generates a design or detail, good or not so good, which immediately leads to alternatives.

So, besides being a delight early design sketches, sometimes very rough, are my way of imparting overall concepts and details to other members of the design team for discussion and modification.

Despite the almost universal use today of CAD (Computer-Aided Drawing), which always gives the impression of a finite final solution, I strongly believe that during the initial phase of any design, hand-drawn sketches are still the best form of communication.

(Preface to *Tony Hunt's Second Sketchbook*)

Sketching always will be an essential part of engineering and architecture. Just believe as John Ruskin did: '*The art of drawing which is of more real importance to the human race than that of writing ... should be taught to every child just as writing is*'.

BIBLIOGRAPHY AND ACKNOWLEDGEMENTS

Amarna 1, Modern Architect's Interpretation of Aspective Plan – Drawing by Mary Hartley after N. de Garis Davies, *The Rock Tombs of El Amarna*. London, 1903, p. 40.

Amarna 1, Aspective Plan of Palace – Drawing by Mary Hartley after N. de Garis Davies, *The Rock Tombs of El Amarna*. London, 1903, pl. XXXII.

Tony Hunt, *Tony Hunt's Second Sketchbook*. Oxford: Architectural Press, 2003.

Andrew Marr, *A Short Book About Drawing*. London: Quadrille, 2013.

Photograph of pond in Rekh-mi-re by Leonie Donovan.

Rehkmire, Two Aspective Ponds – Drawing by Mary Hartley after N. de Garis Davies, *The Tomb of Rekh-mi-Re at Thebes*. Vol. 1. New York, 1953, pl. LXXIX.

D'Arcy Wentworth Thompson, *On Growth and Form*. New York: Dover, 1992 (reprint of revised edition of 1942; first published 1917).

I am particularly grateful to Sir Terry Farrell, Professor Gregory Brooks of the University of Texas, Mary Hartley, Laurie Chetwood, Nina Chislett, Nicola Evans, John Duke and colleagues at WSP | PB and to Routledge editors Fran Ford and Grace Harrison for their encouragement and support during the preparation of this book.

Sketches in Part 2 In addition to acknowledgements provided with particular sketches, other sketches are reproduced by kind permission of Tara Andrews, Gary Boustead, Evelina Gadzhova, John Parker, Alex Lifschutz, Robert Wiesner, Laurie Chetwood, Anderson Inge and the Renzo Piano Building Workshop.

Sketches in Part 3 All of the sketches are taken from work at various engineering/architectural design stages ranging from bid to concept, construction and beyond and were not drawn specifically for this book. They are reproduced by kind permission of WSP | Parsons Brinckerhoff and by kind permission of many of WSP | PB's clients, various institutions and fellow professionals in the world of construction, and include: Chetwood Associates; Zaha Hadid Architects; Cambridge Regional College; The Brit School; Foster + Partners; Sellar Property Group; Mark Weintraub Architects; Masterworks Developments; Heatherwick Studio; KPF Architects; Renzo Piano Building Workshop; Solidere; Farrell Architects; Heron Land Developments; Ronson Capital Partners; Sainsbury Supermarkets Ltd; Mace Group; Knightdragon; RHWL Architects; the Wellcome Trust; University College London; Hammerson plc; Westminster City Council; Will Alsop; St George plc; Liftscutz Davidson Sandilands; Gatwick Airport Ltd; Network Rail; Fulham Football Club; Royal College of Music; Historic Royal Palaces; Howard de Walden Management Ltd; The National Gallery; Canary Wharf Contractors Ltd; Quadrant Estates; Harrods; Treasury and Resources Department | Jersey Property Holdings; Beetham; the Serpentine Gallery; Prime Development; Brookfield Property Partners; Great Portland Estates; Prime Investors; Lady Sainsbury; Royal Mail Group; TH Real Estate; P4P.

INDEX

Page numbers in *italics* refer to figures.

T - #0981 - 101024 - C272 - 219/276/16 [18] - CB - 9781138925403 - Gloss Lamination